Sensing Cities

As cities globally redesign their urban landscapes, they produce a different urban aesthetic and create new experiential milieus. Urban regeneration processes generate radical physical, social and cultural changes in neighbourhoods that demand new conceptual frameworks to address their impact upon daily urban life.

Sensing Cities investigates the reconfiguration of contemporary public space and life through the prism of the senses. The book explores how the increased stylization of cityscapes requires an understanding of public life as a spatial-sensuous encounter. Degen examines how power relations in public spaces are embedded in, exercised and resisted through the sensuous geography of place. This sensory paradigm is then applied to compare two emblematic regeneration projects, namely El Raval in Barcelona and Castlefield in Manchester. By combining detailed ethnographic analysis and interviews with those involved in planning regeneration processes and those experiencing them, the book argues that a changing sensuous landscape is crucial in redefining people's social practices, attachments and experiences in places. Focusing on two European cities at the forefront of urban design, Barcelona and Manchester, Degen draws on sociology, geography, anthropology and cultural and architectural studies to provide a critical account of the politics of publicness in the entrepreneurial city.

With numerous photographs and maps this book stresses the ongoing, embodied and active nature of regeneration as a lived social process rather than merely as a physical or economic exercise. Ultimately, *Sensing Cities* examines how urban regeneration is made effective through the organization of sensory experience. This book is essential reading for students and researchers of architecture, urban studies and human geography.

Mónica Degen is a Lecturer in Sociology at Brunel University where she teaches and researches urban culture, the senses, spatiality and cultural sociology. She has published in *Environment and Planning A*, the *International Journal of Urban and Regional Research* and *Space and Culture*.

Routledge studies in human geography

This series provides a forum for innovative, vibrant, and critical debate within human geography. Titles will reflect the wealth of research which is taking place in this diverse and ever-expanding field. Contributions will be drawn from the main sub-disciplines and from innovative areas of work which have no particular sub-disciplinary allegiances.

Published:

1. **A Geography of Islands**
   Small island insularity
   Stephen A. Royle

2. **Citizenships, Contingency and the Countryside**
   Rights, culture, land and the environment
   Gavin Parker

3. **The Differentiated Countryside**
   Jonathan Murdoch, Philip Lowe, Neil Ward and Terry Marsden

4. **The Human Geography of East Central Europe**
   David Turnock

5. **Imagined Regional Communities**
   Integration and sovereignty in the global south
   James D. Sidaway

6. **Mapping Modernities**
   Geographies of Central and Eastern Europe 1920–2000
   Alan Dingsdale

7. **Rural Poverty**
   Marginalisation and exclusion in Britain and the United States
   Paul Milbourne

8. **Poverty and the Third Way**
   Colin C. Williams and Jan Windebank

9. **Ageing and Place**
   Edited by Gavin J. Andrews and David R. Phillips

10. **Geographies of Commodity Chains**
    Edited by Alex Hughes and Suzanne Reimer

11. **Queering Tourism**
    Paradoxical performances at Gay Pride parades
    Lynda T. Johnston

12. **Cross-Continental Food Chains**
    Edited by Niels Fold and Bill Pritchard

13. Private Cities
Edited by Georg Glasze, Chris Webster and Klaus Frantz

14. Global Geographies of Post Socialist Transition
Tassilo Herrschel

15. Urban Development in Post-Reform China
Fulong Wu, Jiang Xu and Anthony Gar-On Yeh

16. Rural Governance
International perspectives
Edited by Lynda Cheshire, Vaughan Higgins and Geoffrey Lawrence

17. Global Perspectives on Rural Childhood and Youth
Young rural lives
Edited by Ruth Panelli, Samantha Punch and Elsbeth Robson

18. World City Syndrome
Neoliberalism and inequality in Cape Town
David A. McDonald

19. Exploring Post Development
Aram Ziai

20. Family Farms
Harold Brookfield and Helen Parsons

21. China on the Move
Migration, the state, and the household
C. Cindy Fan

22. Participatory Action Research Approaches and Methods
Connecting people, participation and place
Sara Kindon, Rachel Pain and Mike Kesby

23. Time–Space Compression
Historical geographies
Barney Warf

24. Sensing Cities
Regenerating public life in Barcelona and Manchester
Mónica Montserrat Degen

Not yet published:

25. International Migration and Knowledge
Allan Williams and Vladimir Balaz

26. Design Economies and the Changing World Economy
Innovation, production and competitiveness
John Bryson and Grete Rustin

27. Whose Urban Renaissance?
An international comparison of urban regeneration policies
Libby Porter and Katie Shaw

28. Tourism Geography: A New Synthesis
Second edition
Stephen Williams

Sensing Cities

Regenerating public life in Barcelona and Manchester

Mónica Montserrat Degen

LONDON AND NEW YORK

First published 2008
by Routledge
2 Park Square, Milton Park, Abingdon, Oxon OX14 4RN

Simultaneously published in the USA and Canada
by Routledge
270 Madison Ave, New York, NY10016

Routledge is an imprint of the Taylor & Francis Group, an informa business

Typeset in Times New Roman by
Book Now Ltd
Printed and bound in Great Britain by
MPG Books Ltd, Bodmin

British Library Cataloguing in Publication Data
A catalogue record for this book is available from the British Library

Library of Congress Cataloging in Publication Data
Degen, Mónica Montserrat.
Sensing cities: regenerating public life in Barcelona and Manchester / by Mónica Montserrat Degen.
p. cm.
"Simultaneously published in the USA and Canada."
Includes bibliographical references and index.
1. Urban renewal–Spain–Barcelona. 2. Sociology, Urban–Spain–Barcelona. 3. Urban renewal–England–Manchester. 4. Sociology, Urban–England–Manchester. I. Title.

HT178.S62B3724 2008
307.760946′72–dc22 2007049621

ISBN10: 0–415–39799–5 (hbk)
ISBN10: 0–203–89551–7 (ebk)

ISBN13: 978–0–415–39799–5 (hbk)
ISBN13: 978–0–203–89551–1 (ebk)

A mis padres que siempre están conmigo

Contents

List of illustrations xi
Acknowledgements xiii
List of abbreviations xiv

PART I 1

1 Introduction: Sensing cities 3

Urban redevelopment, the senses and public life 7
Researching the senses 12
Overview of the book 13

2 Public life in late modernity 16

Space as social process 17
Sensing publicness 20
A new urban aesthetic 25
Commercial privatization and fortification of urban space 30
Conclusion 34

3 Sensing the city 35

Towards a 'socially embedded aesthetics' 36
The city of senses 41
Place gestures 47
Methodological discussion: rhythmanalysis 50
Conclusion 52

4 Sensuous powers 54

Flows of power 55
Control through pleasure 59
Domination and resistance 62

Sensuous planning ideologies 65
Conclusion 72

PART II 75

5 Castlefield and El Raval 77

A sensuous history 78
Castlefield: from industrial revolution to inner city cool 83
El Raval: from 'Barrio Chino' to cultural quarter 93
Conclusion 103

6 Planning regeneration 105

The spatialization of regeneration 106
Formalizing public life in Castlefield 120
Diluting public life in El Raval 125
Conclusion 130

7 Perceptions from 'down below' 132

Castlefield: exclusive public spaces 133
El Raval: invisible public spaces 139
Castlefield: promoting selective histories 144
El Raval: disrupting social lives 151
Conclusion 161

8 Living in regenerated worlds 162

Rhythms of everyday life 163
Sensuous mappings 173
Castlefield's 'taste wars' 176
El Raval's 'place wars' 179
Conclusion 191

9 Conclusion: Regenerating public life? 194

Notes 201
Bibliography 206
Index 220

Illustrations

Figures

1.1	A street in El Raval	5
1.2	A street in Castlefield	6
5.1	Street map of Castlefield	84
5.2	Context map of Castlefield	85
5.3	'This is not Amsterdam, this is Manchester', Piccadilly Gardens, 2001	90
5.4	Map of Castlefield's Urban Heritage Park	93
5.5	Street map of El Raval	94
5.6	Context map of El Raval	95
5.7	Map of El Raval's regeneration, 2003	100
6.1	Castlefield's Urban Heritage Park in the early 1980s	107
6.2	Castlefield's Urban Heritage Park in 2000	108
6.3	Merchant's Bridge	108
6.4	Promotional leaflet on El Raval's regeneration	111
6.5	Plaça dels Angels around 1900	113
6.6	Plaça dels Angels in 1999	113
6.7	Castlefield's Victorian railway viaducts	114
6.8	Castlefield's Event Arena	115
6.9	Plaça dels Angels, bordered by museums	115
6.10	New and old in El Raval	118
6.11	A traditional street in El Raval	120
6.12	A regenerated street in El Raval	120
6.13	Visitors inspecting a map of Castlefield	122
6.14	An event held in the Event Arena	123
7.1	A small business in the back streets of Liverpool Road	134
7.2	Castlefield's first 'corner shop'	138
7.3	'No flats for sale – OK?'	144
7.4	'Flat for sale'	144
7.5	A sales brochure advertising Castlefield's heritage	145
7.6	The Roman fort replica	148
7.7	Ted and Jerry's newsagent	149

7.8 Leftover walls 152
7.9 An old man observing the demolition of the neighbourhood 154
7.10 Old buildings surrounding the Museum of Contemporary Art in El Raval 156
7.11 'No to the cemetery neighbourhood' graffiti 157
7.12 Old balconies in El Raval 159
7.13 New balconies in El Raval 160
8.1 The bank holiday market in Castlefield 164
8.2 The 'European feel' in Castlefield: Dukes 92 bar 165
8.3 Tourists posing in front of the Museum of Contemporary Art in El Raval 168
8.4 Photo-shoots in El Raval 169
8.5 Photo-shoots in Castlefield 170
8.6 Teenagers appropriating space 171
8.7 Elderly pedestrians resting 171
8.8 Children playing at night on the Plaça dels Angels 172
8.9 Bargains at Castlefield's bank holiday market 178
8.10 'Fed up with the P.E.R.I' 180
8.11 Graffiti being whitewashed 180
8.12 A fashion show on the Plaça dels Angels 181
8.13 A village within the city 183
8.14 An old stationery shop 185
8.15 A Pakistani internet café and call shop 187
8.16 Let's *ravalejar* 192

Tables

5.1 Overview of El Raval's and Castlefield's regeneration 102
8.1 Castlefield 174
8.2 El Raval 174

Acknowledgements

I am indebted to many people in the writing of this book: first, to my mother for teaching me to observe, sense and listen, and to my father for his constant support and belief in me. Thanks to Chris James for his insightful comments and inspiring discussions, for making me laugh and for always being there for me.

I am very grateful for the advice and time of John Urry, Bülent Diken, Deidre Boden, Scott Lash, Malcolm Miles and Rob Shields. I have learnt a lot from them all. Anne Cronin, Monika Buscher, Derek McGhee, Donna Reeve, Javier Caletrio, Sara González, Hester Reeve, Anne-Marie Fortier and Mimi Sheller further provided constant encouragement and feedback during my time at Lancaster University and afterwards. I am indebted to Steve Pile and Gillian Rose of the Open University for reading through and offering advice on the manuscript. Brunel colleagues who have offered advice on diverse chapters include Emma Wainwright, Anna Gough-Yates, Sanjay Sharma, John Roberts and Peter Wilkin. Marisol Garcia, Manuel Delgado and Eugeni Madueño in Barcelona and Michael Hebbert, Dharman Jeyasingham and Tom in Manchester were crucial in opening doors and making the fieldwork a very pleasurable experience. I would also like to thank my friends Nicola Kirkham, Manuela Barz, Jordi Canals and Carmen Petrus for their help. Thanks are also due to Caroline Mallinder, Jennifer Page and Andrew Mould at Routledge.

Above all I want to thank the residents, visitors, planners, architects and politicians in Barcelona and Manchester who I interviewed and without whom this project would have not been possible.

This book developed out of doctoral research at Lancaster University made possible by an ESRC studentship.

The book extends arguments first outlined in the author's papers that were published in *Space and Culture* (vol. 11/12, 2001) and the *International Journal of Urban and Regional Research* (vol. 27 (4), 2003), which are used here with kind permission.

Abbreviations

ARI	Area de Reforma Integrada
CCCB	Centro de Cultural Contemporanea Barcelona
CMDC	Central Manchester Development Corporation
MACBA	Museo de Arte Contemporaneo de Barcelona
PERI	Plan Especial de Reforma Interior
PROCIVESA	Promocio Ciutat Vella S.A.
UDC	Urban Development Corporations

Part I

1 Introduction

Sensing cities

Jane: I like looking at the shapes and how they have changed.
Monica: How have they changed?
Jane: Well, we'd first got dirty scrap-yards and now we've got clean, plain lined, modern buildings.

(Jane, tourist guide in Castlefield)

Monica: And what do you think about the new Plaça dels Angels?
Rafael: A disaster, a disaster as a square, a disaster from an urban planning point of view, and a disaster because of the 'ambience' you get there . . . It is a centre for reunion for all those people: skateboarders, people who take their dogs there to shit, people who just sit around. Not so much in winter but in the summer you get a proliferation of people from the Third World: Filipinos, Pakistanis, Moroccans and they eat there and drink there . . . and throw all the rubbish around.

(Rafael, shop-owner in El Raval)

Walk down any street in any city. What can you hear, feel, smell, taste and see? Your body brushes against other bodies; a cacophony of voices echoes in your ears; your nostrils are teased by smells emanating from shops; your eyes captivated by the bright displays shining through smooth glass windows set against the rough stone of red bricks. Or maybe your body senses the emptiness of a street, perceiving the lack of fellow pedestrians; the overwhelming traffic noise; the taste of smog; the stench of the gutter; and the endless rows of identical housing. Wherever you are in the city, as soon as you move into a street you are involved in an intense sensuous[1] encounter. The senses play a crucial role in mediating and structuring urban experience.

In this book I focus on this important yet largely neglected dimension of urban life: namely the significance of the senses in the (re)configuration of contemporary public space and life. The senses mediate our contact with the world as one engages in public life through the body: 'the city exists in my embodied experience' (Pallasmaa 2005: 40). The architecture, streets, shops and social life we encounter, people's behaviour and activities, all amalgamate through our embodied perception to create a 'sense of place'. The mix of material and social features arouses sensuality and their combination produce the ever fluctuating nature of public life: 'the relationship of public space to public life is dynamic and reciprocal' (Carr *et al.* 1992: 343).

Public places provide the tangible and physical realm for a shared sense of being with other humans. Even in late modernity, with technological advances such as the internet shaping new public spheres, social intercourse in public spaces has not diminished. Indeed, broader social and economic restructurings since the 1970s have meant that the city is the focal point of a new post-industrial economy which attracts ever more people to live in, work in or consume its sites. Significant here is that cities across Western Europe and further afield have been involved during the last two decades in an intensified period of urban renewal that has paid particular attention to the redevelopment and redesign of public space (Harvey 1990, 2000; Hall and Hubbard 1998; Castells 2002; Madanipour 2003, 2006). This remodelling of urban space entails an intense reassembling of the physical landscape that is evaluated in conflicting ways as the above quotations illustrate. My concern in this book is not with charting the reasons and politics behind these processes per se. My concern is rather to understand how regeneration processes transform the sensory qualities of places and whether this sensuous reorganization excludes or includes particular cultural expressions and practices in the public life of these spaces.

To examine the role of the senses in framing public space and life I focus on two neighbourhoods that have witnessed profound transformations in their physical set-up in a relatively short span of time: Castlefield in Manchester and El Raval in Barcelona. Let me briefly elaborate on my reasons for doing so. During the mid-1990s I was travelling regularly between Barcelona, my home town, and Manchester. Spending long periods of time in these cities made me realize that the 'character' of these cities was rapidly transforming as similar changes were taking place. Two areas particularly caught my attention: Castlefield, a place I observed periodically from a railway bridge on my way to Manchester airport, as it slowly metamorphosed from a grey, industrial space into a whitewashed and landscaped leisure area, criss-crossed by canals; then, after a two-hour flight, Barcelona, a city that has reinvented itself dramatically since the death of the dictator Franco in 1975. There, I would spend afternoons walking through the narrow, run-down streets of the lively Raval, situated next to the well-known boulevard of Las Ramblas. Once the infamous red light district of Barcelona, this neighbourhood, like Castlefield, was being 'cleaned up'.

Although the urban landscape and cultures of both cities were very different, similar patterns started to emerge in these neighbourhoods. Both places were reorganizing their urban environment with a special emphasis on redesigning their public spaces. Both areas used the construction of major flagship cultural projects such as the Museum of Contemporary Art of Barcelona in El Raval, and the Museum of Science and Industry in

Castlefield, to inscribe new values into the landscape and promote a new neighbourhood image. This meant that streets were widened, new street-furniture was introduced, buildings were sand-blasted, and so on. Indeed, these new public places *felt* and *looked* remarkably similar.

However, the existing physical set-up of each place and their socio-cultural setting could not have been more different. El Raval contains a varied building stock ranging from a Romanesque church to fifteenth-century buildings and predominantly nineteenth-century housing. It is an area characterized by confined spaces, a network of small squares, narrow streets, courtyards and passageways. Historically a neighbourhood of crowded living conditions, it was still densely populated at the time of its regeneration. Many of its residents have lived here for two or three generations, a feature reflected in the large number of family-run corner stores and workshops on its streets and its intense vernacular street-life. By contrast, in Castlefield the existing building structures were mainly those of

Figure 1.1 A street in El Raval. (Photograph by author.)

Figure 1.2 A street in Castlefield. (Photograph by author.)

a nineteenth-century industrial landscape, with large open spaces interwoven with canals and hardly any resident population to take into account at the start of the regeneration. Despite contrasting social environments – a southern European neighbourhood with an already rich public life, and a northern European one, lacking public life – a similar form of global spatial practice was enforced to change the existing public space: cultural regeneration (Bianchini and Parkinson 1993).

Urban regeneration schemes are long-term projects dependent on cycles of investment that make it difficult to assess when a place is 'finished'. Both Castlefield and El Raval are constantly evolving. Indeed, Castlefield is now populated by a largely homogeneous, predominantly white, middle-class population, whereas El Raval has been transformed into 'Barcelona's cultural paella pan' (Time Out 2001: 76), and is one of the city's most racially mixed neighbourhoods. I have therefore restricted the focus of my empirical study to the years between 1997 and 2002 when the key spatial restructuring took place and the socio-spatial and sensuous changes were most acute and perceptible. In exploring the links between spatial transformations and sensuous experience the book provides a critical account of the impact of regeneration strategies on the experiential landscape of cities. By way of introduction I elaborate briefly on how the three key themes that run through my book – globalization and urban change; the senses and urban experience; and public space and life – are connected, and set out why a focus on the senses is necessary to examine contemporary processes of urban redevelopment.

Urban redevelopment, the senses and public life

In her article assessing the development of urban sociology at the start of the new millennium the sociologist Saskia Sassen states that after a decline of interest in the city during the second half of the twentieth century, 'the city is once again emerging as a strategic site for understanding major new trends that are reconfiguring the social order' (2000: 143). The key new trend that Sassen refers to is globalization. Her argument is that with the decreasing role of the nation state, cities are becoming the main players in the world economy and are using their physical space to attract business, tourists and global investment. In this global network, places become interrelated structurally through a range of features such as international economic and political networks; an increasingly mobile 'public' constituted equally by tourists, commuters or migrant populations; or through the exchange of ideas and discourses. This intensified cross-reference between places fostered by global processes has meant that the notion of 'place' as a fixed and bounded entity has undergone a number of critical re-evaluations (Keith and Pile 1993; Eade 1997; Massey 1995b, 2005). Place is increasingly understood 'as a meeting place, the location of the intersections of particular bundles of activity spaces, of connections and interrelations, of influences and movements' (Massey 1995b: 59). Such a 'dialogic' notion of place acknowledges the inherent complexity of place materially, socially, politically and economically, and views globalization as far from a unified, singular process, but rather as multiple and unstable (see King 2004 for a overview of the use of this concept). In other words, globalization processes are not uniform but place-specific, as they get reworked in and through particular local contexts. Cities are precisely the places 'where a multiplicity of globalization processes assume concrete, localized forms . . . Recovering place means recovering the multiplicity of presences in this landscape' (Sassen 2000: 147).

Urban regeneration schemes present ideal sites to analyse the 'multiplicity of presences' and to examine how these diverse and often contradicting processes of globalization are transformed and appropriated in different locations by diverse audiences. Or, to put it more poignantly, to help us realize that '[w]hat is seen as globalization looks very different from different points of view' (King 2004: 42). Let me briefly explain this point by referring to Baltimore's harbour redevelopment which can be regarded as the template for many regeneration projects both in the United States and Europe. The transformation in the early 1980s of a once industrial area filled with disused wharves, warehouses and railways into a refurbished retail and entertainment festival market would become a much cloned model all over the globe, and regarded as financially beneficial to the image and economy of the city. One of the principal aesthetic features of regeneration projects is the physical transformation of the landscape through the integration of heritage, landscaping and leisure in the design of place, both in order improve the 'quality of life' in the area and to attract visitors. As Harvey explains, functionalist architecture was replaced in Baltimore by 'an architecture of spectacle, with its sense of surface glitter and transitory participatory pleasure, of display and ephemerality, of jouissance, [that] became essential to [its] success' (1990: 91). Significantly a key element in the successful redevelopment of Baltimore was the alteration of the sensory properties of its material landscape.

For some the regeneration of Baltimore has been successful because the remodelled space has attracted new business and visitors and provided an attractive image of the city that has fostered investment and civic pride. Critics, however, point out that such regenerated landscapes are 'pockets of revitalization surrounded by areas of extreme poverty' (Levine quoted in Hannigan 1998: 53). As Levine (1987), and more recently Harvey (2000), argue in their assessment of Baltimore's redevelopment, the local population benefit little from these schemes, and are often vehemently opposed to them. The reasons range from more structural arguments, such as the worsening of the social and economic conditions of the less well-off due to spending on prestige areas taking precedence over welfare services, to a painfully personal perceived erasure of local culture reflected in the demolition of familiar sites and buildings. Similar criticisms have been voiced against the regeneration of Birmingham (Hubbard 1996), the London docks (Foster 1999), Manchester (Quilley 2000) and Barcelona (Heeren 2002; Degen and Garcia 2008). Hence, as a global process, regeneration is perceived very differently by a range of social groups affected and involved in these processes. Yet little attention has been given to local variations or the lived culture of entrepreneurial cities (Hall and Hubbard 1998), and few studies have analysed how these new landscapes are embedded in the everyday life of a range of social groups.

To date, most studies of the cultural transformation of urban space have analysed three main themes: the privatization and commodification of public space (Sorkin 1992a; Hannigan 1998); the rising influence of culture in determining the cities' economy (Zukin 1995; Scott 2000); and gentrification (Ley 1996; Smith 1996; Atkinson and Bridge 2005). What these studies have in common is that they highlight how culture has become a key factor in the development of urban planning considerations, in the negotiations around the use of public space, and in the shaping of public life. Culture is the new battleground in cities which determines who is included or excluded in the public life of urban environments (Zukin 1995; 1998a). This book takes these positions as a starting point to expand the study of urban change in a number of new directions.

First, most accounts of urban change tend to focus narrowly on the American experience of major global cities and examine mainly large-scale urban developments[2] (Soja 2000). However, with the expansion of neo-liberal urban politics across the globe these features are also to be found in the regeneration strategies of many neighbourhoods, both in Europe and elsewhere (Peck and Tickel 2002). The novelty of new global urban conditions such as regenerated spaces provides a rare opportunity for comparative cross-cultural research on the discourses that inform the planning of these spaces, the practices of their users[3] and the social significance that regenerated public space has in the entrepreneurial city. There is a general lack of comparative analysis on the restructuring of urban space from a cultural perspective that interrogates the 'localism' of place and considers how different local circumstances and a different set of local attitudes might reframe the outcomes of global cultural processes. In what ways does the vernacular culture play itself out within similar regeneration processes? What do these transformations mean for the type of public life that is created in regenerated urban environments?

Second, the experience of cultural and physical space restructuring tends to be summed up mainly from a visual perspective, often under the concept of the 'aestheticization of everyday life' (Featherstone 1991), or 'spectacularization' (Boyer 1988, 1992, 1995; Hannigan 1998). These environments create a distinct aesthetic in their spatial arrangements, their promotional campaigns and their sensuous perception, creating new experiential landscapes. Yet the changing urban space is not experienced via the visual sense alone but through the whole sensory body. A central issue in this book is therefore to understand how the sensuous perception of social space contributes to or undermines the social and ideological cohesion of the city. In other words, how does the sensuous organization of public space enhance or exclude the participation of particular groups in public life?

Third, studies on gentrification have tended to demonstrate how neighbourhoods transform socially through the influx of the middle class and the gradual expulsion of working-class or ethnic minorities. I broaden this analysis by considering further factors of neighbourhood change, particularly those that transform the individual's attachment to place. A sensuous-spatial analysis of neighbourhood change provides a framework to analyse the relationship between physical, cultural and social transformations of public place. While literature on the spatial addresses the influence of capitalism in defining power structures in spaces (Harvey 1990; Castells 1996; Zukin 1991) and the imposition of landscapes and aesthetic taste (Zukin

1988; Featherstone 1991), little research has been done into how this affects daily social interactions: in other words, quotidian public life in these landscapes. My study aims to get closer to the lived and embodied experience of the agents involved in and affected by urban restructuring. What are the underlying cultural codes that guide the reorganization of certain neighbourhoods? How do different groups negotiate these? What is the nature and articulation of urban change within lived embodied experiences?

In order to link transformations of space with an analysis of sensuous experience this study follows the tradition of thought initiated by Lefebvre (1991), who began to look beyond space as a 'container' for social action and instead started to interpret space as a product and producer of multiple forms of spatial practice: 'Space is permeated with social relations; it is not only supported by social relations but it is also producing and produced by social relations' (Lefebvre 1991: 286). Space is thus always political, meaning that social power relations are expressed in and through space. A focus on the senses offers a way of analysing the relationship between built form and social relations, as senses provide the framing texture for the material and social bond in public spaces. Paying attention to how the senses frame our experience of cities invites us to capture a largely ignored aspect of city life that is as significant as their physical structure, namely their 'character' or 'mood'. Steve Pile makes a strong case for analysing more than the physical form of the city and for focusing on the less tangible features that are equally important in defining the essence of the city: 'what is real about cities, then, is also their intangible qualities: their atmosphere, their personalities, perhaps' (2005: 2).

In the first part of the book I therefore develop a theoretical framework around the concept of 'socially embedded aesthetics' that allows me to analyse how social relations in public spaces are constituted by, exercised through and embedded in the sensuous geography of place. I contribute to a growing amount of research on the senses and society. While there has been an increasing interest in the senses in recent years (Urry 1999, 2000; Howes 2005a; Bull *et al.* 2006), few studies have focused their attention on the ideological importance of the senses in the restructuring of contemporary urban life (exceptions are Law 2001; Bull 2000; Howes 2005b). As Howes poignantly comments: '[it] is not only a matter of playing up the body and the senses through evocative accounts of corporeal life, although these can be valuable, but of analysing the social ideologies conveyed through sensory values and practices" (Howes 2005a: 4). Examining the sensory politics in urban environments helps to reveal more subtle forms of power that are transmitted and experienced through cultural practices.

The final point to tackle with regard to the arguments made in this book is the 'end of public space' thesis (Sennett 1986; Brill 1989; Sorkin 1992b; Kilian 1998). The best-known argument is Sennett's (1986) concern about the widening gap between public and private experiences which affects how we relate to others in public spaces. Sennett's thesis is complex, and I do not have the space here to fully engage with his arguments, but I would like to point out the key issues that concern my study. According to Sennett, during the eighteenth century cities such as London or Paris expanded and developed networks of sociability independent of royal or feudal control 'where strangers regularly meet' (1986: 17). These new spaces of sociability were promoted through the building of urban parks, the creation of broad avenues for strolling and the opening of theatres, coffee-houses and coaching inns for the general public. Yet, as capitalism and secularism took hold in major Western cities during the nineteenth century, and unsettled the existing social order, the bourgeoisie gradually retreated from public life to seek refuge in the family, and encounters with strangers in the public realm were regarded with increasing distrust, or even actively shunned. This erasure of public life has been further exacerbated by the rise of modernist architecture and urban design in the twentieth century. Modernist architecture combines an aesthetic of visibility and social isolation to produce spaces that lack a diversity of activities and discourage the passerby from sitting and spending time, and 'as public space becomes a function of motion, it loses any experiential meaning of its own' (Sennett 1986: 14). The rise of car ownership and our dependence upon it has meant that twentieth-century Western urban planning has largely approached public space as a vehicle for traffic and motion, and this has led to a further reduction of public space. The last post for public space was effectively sounded in the 1990s when critics identified an increasing privatization and commercialization of public space and denounced a further erosion of the public sphere (Mitchell 1995, 2003; Boyer 1993; Sorkin 1992b; Fyfe 1998).

While I am sympathetic to some of these claims, two important limitations need to be pointed out. First, public sphere is a very elusive concept 'defined in terms of processes and dynamics rather than institutions or geographical borders in which citizens have an incentive to lay aside particular interests to adopt a public interest perspective' (Garcia 2006: 752). This means that a reduction of public space does not naturally lead to a reduction of the public sphere; in fact it has been argued that through new communication technologies the public sphere has been expanded. New public forums and forms of citizenship are emerging beyond state-defined locations, for example in environmental activism or AIDS activism (Brown 1997; Stevenson 2005).

Second, the 'end of public space' arguments seem to imply that there was a time when public space was characterized by diversity and accessibility for all. However, as feminist writers have pointed out (Fraser 1990; Wilson 1991; Duncan 1996; Deutsche 1996), these accounts are based on an idealized conception of public space that has in reality never existed. Public spaces are highly contested and exclusionary spaces. Public space has always provided open and lively places for some – predominantly white males – and 'dead' or threatening spaces for others – women, ethnic minorities or those that do not conform to mainstream expectations. This suggests that the designation of a place as public or private is not established through physical boundaries alone but rather that social conventions and practices determine who and what is allowed to be 'public'. Public space is hence a historically and socially contingent notion. Indeed, it might be 'more useful to imagine public space as constituted by difference and inherently unstable and fluid' (Bridge and Watson 2000: 374). As this book argues, public and private spaces are better thought of as sensory-qualitative experiences, bounded through spatial and temporal dimensions that are produced through a range of practices and conventions and are therefore always open to manipulation and change.[4]

Researching the senses

As a cultural sociologist fascinated by cities, I am interested in how cities retain their unique character and identity despite an intensification of globalizing processes and claims of standardization in urban life (Sudjic 1992). An inherent problem of researching public life is that it is made out of an amalgamation of overlapping personal and institutional spatialities and temporalities. For this reason I have decided to focus on and foreground the various layers that form urban social space, such as design, representation, use and experience (see also Low 1997). The outcome has been a variety of intersecting, sometimes contradictory, stories: my own observations and interpretations, the diverse experiences of interviewees, and finally the story that the images in this book tell. Taken together they are intended to provide their own evocation of the senses, and of the embodied and very often ephemeral qualities of place.

To analyse the connections and relations amongst geographically distant places I decided to follow the suggestion of Burawoy *et al.* to develop a 'global ethnography' which 'entail[s] a shift from studying "sites" to studying "fields", that is the relations between sites' (2000: xii). Such a comparative ethnography provided me with a method of understanding and comparing the ways in which different social groups involved or affected by the regeneration process have experienced the transformations of public space and life on a daily basis. My daily presence over eight months[5] in El Raval and Castlefield as a 'participant pedestrian' allowed me to become acquainted with a diversity of users, from long-term residents, shop-owners and commuters to tourists, night revellers and prostitutes, whom I gradually approached and interviewed.[6] My aim was to get an insight into the diversity of experiences and reactions to the regeneration of the neighbourhoods. In order to understand the reasoning behind the spatial restructuring and transformation of public life I also interviewed urban planners, politicians and architects involved in the regeneration strategies. In each case my priority was to get the interviewees to determine the interview agenda, giving priority to their 'sensing' of the regeneration.

Interviews and official documentation were analysed not only in terms of what was said, but also in terms of what was left unsaid and suggested (Bourdieu 1999). When talking to people about the character and sensory make-up of places there was always something that could not be quantified merely through words or descriptions, such as the 'feeling' of a place, the lived public life, and memories of places filled with emotions, only perceptible in the trembling of a voice or silent sight. For example, one elderly gentleman in El Raval started to hum a childhood song when I asked him what had disappeared from the neighbourhood – how are we to interpret this? Similarly, in official documents and representations the sensuous discourses which underpin regeneration processes are not explicitly remarked upon, but rather suggested through their absence. Certain senses are not mentioned or are only evoked by non-existence, such as a house which has been demolished or the replacement of one set of activities in space with another. Consequently, one should read document and interview quotes in this book as 'translations' (Bourdieu 1999) in which I rewrite and embed individual experiences and descriptions within my interpretative framework. Following Clifford's claim that ethnographies are fictions and '[e]thnographic truths are . . . inherently *partial* – committed and incomplete' (1986: 7, emphasis in original), I cannot aim to offer an overarching vision of the sensuous implications of regeneration projects, but instead confront the different views and representation to 'work . . . with the multiple perspectives that correspond to the multiplicity of coexisting, and sometimes directly competing points of view' (Bourdieu 1999: 3).

Overview of the book

The following chapters discuss the regeneration and changing public life of two neighbourhoods through a sensuous prism. The chapters are organized into two parts, the first of which is largely theoretical and develops a framework to analyse the changing politics of public space and life through the senses. By drawing on a diverse range of urban theories from sociology, anthropology, architecture and planning, I illustrate how the senses have been crucial in shaping the form and experience of the city. Those readers more interested in the actual case studies should turn to the second part of the book in which I provide an empirical analysis of the changing sensuous geographies and public life of El Raval and Castlefield.

Chapter 2 starts with an assessment of which sensuous-spatial qualities confer the experience of 'publicness'. Drawing on Lefebvre's notion of spaces being socially produced and experienced through a sensuous body, I develop three dimensions against which we can assess the publicness of spaces: the economy of access, the ethics of engagement and the politics of representation. Such an analysis becomes increasingly pressing as a new urban aesthetic is generated through urban regeneration schemes that change the material and symbolic texture of places.

Chapter 3 provides a theoretical framework for conceptualizing the sensing city. Drawing on the original Greek meaning of aesthetics as sensuous experience, I develop the term 'socially embedded aesthetics' to capture the multi-sensory aspects involved in the physical transformation of public places. I explain the ways in which the senses organize social and cultural spatialization, by which I mean the ways through which space is always culturally coded and therefore ideological.

Chapter 4 interrogates how power is expressed in public space through its sensuous organization. Elaborating the argument that contemporary society has moved from Foucault's 'discipline society' to a 'society of control', I suggest a fluid and relational notion of power. In this context I consider how the production of urban forms has historically been shaped by sensory regimes, and show how these reveal the roots of contemporary sensuous paradigms in regenerated public places in the form of ideologies that aim to control disorder, impurity and exposure to the stranger. Although dominant sensuous experiences can be rooted in the sensuous landscape, the potential for resistance can be equally ingrained and enacted through the sensuous body.

In the second section of the book, I apply this theoretical framework to the two neighbourhoods under study. Chapter 5 provides a passage to the case studies by contextualizing the sensuous transformation of both neighbourhoods' public places in time and space. To understand the spatial, material and symbolic transformations of El Raval and Castlefield, a comparative sensuous history of both places is provided, followed by an overview and analysis of recent urban regeneration policies. In this way, the redevelopment of these neighbourhoods is embedded within broader discussions of recent spatial transformations in Barcelona and Manchester.

Chapters 6, 7 and 8 should be read in relation to each other. Drawing from interviews with the variety of actors involved in the 'making' of these places, ethnographic observations, documents and photographs, the chapters analyse the sensuous-social production of regenerated public space in the case study areas. Chapter 6 starts by examining the 'spatialization of regeneration' in Castlefield and El Raval as it is conceived of by official agents together with their motivations. In this context I examine three aspects: first, the physical reorganization of the neighbourhood shaped by the strategy of accessibility and 'designer heritage aesthetic'; second, the role that regenerated public places are expected to perform within each neighbourhood; and, third, the envisaged public life that the regeneration is expected to foster in each area.

Chapters 7 and 8 focus on the various users' experiences of public space and their evaluation of the regeneration process. Chapter 7 analyses the responses to and perceptions of the officials' spatial strategies of those who use the spaces. I examine how representations of places inform lived experience by analysing the competing and divergent responses to the transformations. This chapter highlights the various forms of attachment to and experience of place, as they are perceived by different groups such as tourists, new and established residents, shop-owners, and so forth, showing how the users actively reproduce and at the same time contest the official strategies.

Chapter 8 explores the lived experience of these regenerated neighbourhoods. Following Lefebvre (1991), I show how everyday life, as experienced through the sensuous body and in terms of activity rhythms, evades hegemonic experiences and practices. By analysing both the spatial rhythms of the areas and the sensuous maps of users I discuss how the transformations in the environment are negotiated through daily consumption of the spaces. In both areas the response to the abstract process of regeneration is shaped by concrete claims of social groups upon public space and expressed in the sensuous conflicts of daily life.

Chapter 9 consolidates the arguments of the book and reflects on the theoretical and empirical implications of the themes developed, offering recommendations as to how 'urban regeneration' might be better effected in other places.

2 Public life in late modernity

Western city centres have experienced an unprecedented revival in the last decade. After major disinvestment in European cities during the 1970s, the late 1980s witnessed the beginning of a radical redesign, refurbishment and renewal of the urban landscape. Formerly empty city centres such as Manchester have seen their population grow by more than 15,000 residents between 1995 and 2007. During the same time property prices of already densely populated urban cores, such as Barcelona's Old Town, have increased by over 250 per cent (Marshall 2004). Not only has the city become 'trendy' (Jencks 1996), but city life has been reinvented or rediscovered, as the following newspaper extract suggests:

> [Cities] are becoming fashionable again. People are trickling back to once-seedy, depopulated areas, where converted warehouses – the ubiquitous lofts – and new flats provide attractive homes. Restaurants, bars and cafe societies have sprouted everywhere. High streets are reviving as city retailers slowly win the battle against out of town shopping centres.
>
> (*The Guardian*, 13 September 2000)

The physical reorganization of public space is celebrated by some as an energizing force for a post-industrial urban economy and public life. Simultaneously, it is criticized by others as aiding the expansion of commercial interests into public space and impoverishing its public sphere. Whichever position one takes, these trends have reinvigorated academic discussion about what 'public space' is and how it is experienced. This book adds to the debate by arguing that as cities around the globe are redesigning their urban landscapes, they are generating a new urban aesthetic which in turn is producing novel experiential milieus. This has important consequences for urban life as new environments reconfigure and restructure patterns of sociality within the city.

By drawing on Lefebvre's (1991) understanding of space as socially produced and sensuously constituted this chapter explores what it is that comprises the actual 'publicness' of places, i.e. the nature of places being public. I continue by outlining the emergence of a new self-conscious urban aesthetic fuelled by heightened pressure on cities to redevelop their environments in light of an intensified global competition to sell specific place experiences. The chapter ends by suggesting that these environments foster new forms of sociability organized around increasingly commercially regulated forms of public life.

Space as social process

During my first days of research in El Raval I was walking along one of its narrow winding streets reflecting on the purpose of my study. It was a busy street with elderly Spanish women brushing past, pulling their shopping trolleys across the pavement, young Pakistani men congregating in corner-shops chatting loudly, and screaming children running past me to school. A window display in an old shoe shop caught my attention. Yellow paper star signs with *oferta* (sale) written on them were advertising a cluttered display of old-fashioned sandals and a pair of Nikes bleached by the sun. Then, in the corner of the window, I read a sign: *Aqui hay vida!* (There is life here!). These words puzzled me. I looked at my surroundings more carefully. Many of the shop-shutters were down, some of them with signs saying: 'We are closed due to the regeneration of the neighbourhood. We advise you to shop elsewhere.' A few doors down, a sleek new pastel-coloured building contrasted with the time-ridden elaborately designed facades of its neighbouring nineteenth-century buildings. On its ground floor a bakery chain had just opened, the aroma of fresh bread mixing with the otherwise musty odours of El Raval, and soothing jazz music was spilling onto the street. Inside, some of its metal designer stools were still covered in plastic, and young men and women in suits (I guessed they were staff from the recently opened museum and art galleries) were drinking coffee.

Months later I was interviewing Sarah, a resident in Castlefield. Since its redevelopment in the early 1990s the area had changed gradually from an empty, derelict space to one that attracts large crowds of visitors and tourists. While Sarah enjoyed living there, she did mind the noise of the café-bars at weekends and the lack of shops for residents and concluded:

> [Castlefield means] home. But, you know I don't have control over my own backyard basically. Things are done to me whether I like them or I don't. But, I'm quite proud of it as an area for development . . . The concern is how it's managed, not just the fabric of it, but when they decide to stop and what the balance is. Because I think it's a pity if it all begins to look too twee, too sanitized . . . There's a difference between it being run-down and derelict and so much just geared for tourism.
>
> (Sarah, long-term resident)

Both descriptions highlight the importance of sensory perception in our relationship to public life in cities. They illustrate a stark sensory contrast between old and new, and in the ways various social groups invoke and produce different sensory practices. In fact we can observe in these descriptions how physical layout, social life and political dimensions are linked in the experience of public life.

Let me briefly elaborate on this point. Physical changes in the urban environment create new patterns of sociability as they attract different spatial practices and social groups. The transformation of a former industrial area into a heritage park, as in the case of Castlefield, involves closing down certain businesses such as car-repair or scrap-metal workshops that are regarded as unsightly. To attract a new public, new venues such as bars and museums need to be opened to provide an incentive for people to walk around, gaze, and relax with a coffee in the new heritage park. In other words, the physical structure of places influences *how* this space is used and *who* uses it. The physical organization of places and the uses of space promoted reflect an important political dimension of the social life in cities as public space operates as 'a *space for representation*, a place in which groups and individuals can make themselves visible' (Mitchell 2003: 33). In the case of El Raval, one of the ways in which a long-term shop-owner chooses to voice his opposition to the regeneration is by visually stating his right to be there, his right to the city. In Castlefield, Sarah does not feel that her needs and rights to place as a resident are reflected in the refurbishment of the area. What is at stake in these contested views and practices of space stems from the clashing of two opposite spatial dimensions: the conceived vision by planners and politicians of what constitutes 'appropriate' activities and sensory experiences in public space and the actual lived practice of place by those using it (see also Mitchell 1995).

To understand how these contestations around place are produced, and to clarify some of the complexities between the lived and the conceived city, let me recapitulate Lefebvre's widely discussed triad of social space (Soja 1996, 2000; Shields 1989). We need to do this, as a central issue in this book is Lefebvre's theorization of social space in relation to what he terms the 'practico-sensory body'. In Lefebvre's view we experience space first and foremost through the sensuous body. Hence, theorizing space involves a focus 'on the body's implication in and constitution of a "sensory-sensual space"' (Simonsen 2005: 1). The body can be understood as a mediator of the coexisting, concording and interfering relationship between the three elements that constitute the social production of space: spatial practices, representations of space and spaces of representation (Simonsen 2005). I briefly outline below Lefebvre's three moments of spatiality, indicating their relevance for a sensory analysis of space.

Spatial practices encompass the daily comings and goings of people, their *perceived* social relationships, the actual physical arrangements of objects in space; the immediately observable and measurable sensory qualities of spaces; in other words, that which can be directly seen, heard, smelled, touched and tasted. It involves a cohesive spatial performance, often played out routinely over time, a taken-for-granted and unreflective practice (Shields 1989). As Soja notes, it is 'the process of producing the material form of social spatiality, [it] is thus presented as both medium and outcome of human activity, behaviour, and experience' (1996: 66).

Representations of space refers to the space of scientists, technocrats and planners. It is the rationally abstracted space which is mentally *conceived* of in verbal, visual or written representations – 'the knowledge of this material reality is comprehended essentially through thought, as *res cogito*, literally "thought things"' (Soja 1996: 79). Conceived space is linked to the relations of production and to the order or design that these produce, and 'are thus the representations of power and ideology, of control and surveillance' (Soja 1996: 67). Conceived space plans, fosters or imposes certain sensory perceptions. It tries to define the ways in which a place will be felt and experienced by its users, from the texture of the pavement to the spatial movements of cars and pedestrians. As Lefebvre argues, spatial practices and representations of space cannot be separated, but have to be thought of as interpenetrating. Nevertheless 'conceived' space has the tendency to dominate 'perceived' space, especially in urban theory, which leads to a neglect of the lived practices and its experience in much academic writing. Yet,

> spatial practice is lived directly before it is conceptualized; but the speculative primacy of the conceived over the lived causes practice to disappear along with life, and so does very little justice to the 'unconscious' level of lived experience per se.
>
> (Lefebvre 1991: 34)

To capture this more personal, direct relation to place, Lefebvre suggests a third moment in the production of social space, namely *spaces of representation.*[1] This refers to the ways in which space is directly *lived* through and moulded by everyday actions, memories, experiences and feelings towards space by its inhabitants and users. It refers to the 'imaginary geographies' that people actively create through their subjective or personal involvements with places. It is 'the dominated – and hence passively experienced – space which the imagination seeks to change and appropriate. It overlays physical space, making symbolic use of its objects' (Lefebvre 1991: 39). It refers to the personal associations and values that individuals associate with particular place experiences. The lived is a unifying place, which encompasses the conceived and perceived through personal perceptions and relations. It is the space in which the micro-relations of power are played out in everyday life, the tangible space of domination and resistance.

All three moments of the above trialectic relate to each other; together they create 'social space'. Each moment in space is informed by and constituted through the other two spaces. Spatial practice, representations of space and spaces of representation are best conceived of as analytical spaces that contribute differentially to the production of place, varying according to local conditions (Dear 2000) and fusing with different intensities. By constructing this triad as a series of dialectic relations, Lefebvre brings together the physical aspect of space as well as attitudes and habitual practices, which leads Shields to suggest that Lefebvre's notion of space 'might be better understood as the spatialization of social order' (Shields 1999: 155). The term 'spatialization' brings us closer to Lefebvre's thinking about space as a social process, and as something that is constantly being made and re-made.

Sensing publicness

In regard to our initial question of how to assess the publicness of an urban place, Lefebvre's conception of space as an ever evolving process and as socially produced supports the notion of public/private space as historically and culturally contingent. Indeed, critics such as Weintraub (1997) and Sheller and Urry (2003) point out that inherent in the various conceptions of public and private (such as public/private interest; public/private sphere; public/private life; public/private space or publicity/privacy) are a variety of meanings and uses shaped by different social science traditions. Common to them all is that they tend to rely on binary categorization of public/private and ignore overlaps or threshold spaces in which public/private tensions are played out together. Yet urban environments are most of the time composed of spaces with different degrees of publicness and privateness (Akkar 2005). Sheller and Urry therefore suggest a conceptual move from 'spaces' to 'moments' in which publicness or privateness might arise, so that we can capture 'the multiple mobile relationships between them, relationships that involve the complex, fluid hybridizing of public-and-private life' (2003: 108). I suggest that one way of capturing the fluid nature of public/private moments is to focus on how they are *experienced* in physical public space. Thus, how do we actually come to perceive a particular space as public? I contend that by analysing the relationship between the sensory-spatial structure of places and the degrees of publicness that these spaces allow for spatial use and imagination, we can start measuring and understanding the experiential qualities of public and private life. Historically three dimensions can be identified as shaping the sensuous experience of publicness: what I define as the economy of access (the physical dimension), the ethics of engagement (the social dimension), and the politics of representation (the political dimension) which I discuss for the remaining of this section.

The economy of access

Strictly speaking public spaces are 'those areas of a city to which, in the main, all persons have *legal access* . . . Public space may be distinguished from private space in that access to the latter may be *legally restricted*' (Lofland 1973: 19, emphasis in original). Hence, the possibility or denial of physical access is a crucial feature in defining the public nature of particular places.

Historically, the notion of public space can be traced back to the Greek 'agora'. This was an open space situated in the centre of a city or, in port cities, along the harbour, often surrounded by civic buildings and serving a variety of purposes from market place to a forum for political discussion, and thus a place of constant social interaction (Zucker 1966). The agora was an open, fluid space, part of the existing road systems and easily accessible for a broad range of people (Wycherly 1962: 50). It served as a centre of communication 'where the interchange of news and opinions . . . play[ed] almost as important a part as the interchange of goods' (Mumford 1961: 149). It was a place for direct and unmediated transactions and stood in stark contrast to the official political function of the acropolis which was legally restricted to the 'legitimate' citizens of the polis, male property owners.

The agora's public character evolved out of its role as a threshold space in which boundaries between the public and private sphere, public and private activities were constantly shifting and subverted by the free movement, discourse and engagement of people. The notion of citizenship or political rights in those times and still nowadays is precisely related to this *economy of access*[2] that public places offer as spaces for potential political representation in the form of a forum that facilitates gathering, talk and visibility for a range of individuals.[3]

A theme that crystallizes from the above discussion is the crucial role that sensuous experience plays in fostering this sense of 'publicness'. The perceptible openness fostered by the agora's design, and its role as a meeting place for a diversity of activities and people, is precisely what establishes it as a place where people can interact as equals. As critics such as Sennett (1994) and Jacobs (1961) highlight, it is the free movement across cities, the access to a diversity of sensory experiences that one engages with: the smelling, touching, seeing, hearing, tasting; the sensuous collisions which create publicness. However, as both critics argue, modern orthodox planning with its principles of segregation, control and order of spatial activities and sensory perceptions has had a tendency to reduce the potential of public space to foster diversity in urban experience and social relations.

The importance of the senses in our engagement in and with public space becomes further exposed when we conceive of being in public from a phenomenological perspective as 'dwelling' in public. In his pivotal article 'Building, Dwelling, Thinking' (1977), Heidegger examined the etymological roots of the German word for building, *bauen*, and found in the concept an essentially experiential echo, as building originally means to dwell: 'The way which you are and I am, the manner in which we humans are on earth, is "buan", dwelling' (1977: 325) Heidegger's phenomenological understanding of being in place highlights the importance of embodied experience in the 'making' of place. Whether one is a resident or passerby, one momentarily dwells in public space, sensuously digesting the place as we personalize a particular location, whether public or private, by reading and interpreting environmental clues through sensory interaction.

In addition to considering the role of physical access, attention has to be paid to the role of symbolic access in the construction of publicness. The access of public space is often restricted to expectations of 'proper' behaviour. For example, displays such as washing in a public fountain are regarded as unsuitable in Western public space and might be legally penalized. Places and their attached meanings are products of certain forms of rules and activities that are culturally specific. Thus, '[t]he effect of place is not simply a geographical matter. It always intersects with socio-cultural expectations' (Cresswell 1996: 8). As I show in the reminder of this book, sensuous regimes play an important role in defining these socio-cultural expectations as they shape who or what is included or excluded in public space and life.

The ethics of engagement

City life is characterized by 'a world of strangers' (Lofland 1973). To be able to live amongst unknown others and minimize the feelings of uncertainty and fear that this might evoke, urban dwellers resort to a range of social and spatial practices to negotiate daily encounters with 'Others' in public space. Simmel refers to these encounters as the *Spielform* (playform) of social life where people give up part of their self, for example their social and cultural status, to participate in the public sphere, where 'one "acts" as though all were equal' (Simmel 1971a: 133). Thus individuals come together in the crowd with a 'sense of toleration of differences' (Wirth 1995: 71). In Simmel's view we are drawn to take part in these forms of urban sociality because it helps individuals to overcome their ontological sense of loneliness in the world: 'above and beyond their special content, all these associations are accompanied by a feeling for, by a satisfaction in, the very fact that one is associated with others and that solitariness of individuals is resolved in togetherness, a union with others' (Simmel 1971a: 128). To put it simply, people sometimes enjoy the presence of others – while at other times they cannot wait to get away from the crowd (see also Tonkiss 2005). While I do not want to romanticize this relationship to the Other, which is certainly a normative ideal, it is based on an careful negotiation of proximity and distance, which I refer to as an *ethics of engagement.*[4] The negotiation of this ethics of engagement is described by Sennett (1986) as 'civility' through which individuals distance themselves internally from the encounter with the Other, and retain a certain level of privacy. Civility is:

> the activity which protects people from each other and yet allows them to enjoy each other's company. Wearing a mask is the essence of civility. Masks permit pure sociability, detached from the circumstances of

> power, malaise, and private feeling of those who wear them. Civility has as its aim the shielding of others from being burdened with oneself.
>
> (Sennett 1986: 264)

While civility in Sennett's account is very positive, aiding social relations, Brewer (1997) highlights in his account of eighteenth-century parks how forms of behaviour such as civility were used to exclude the plebeian class from new bourgeois urban spaces. Thus civility needs to be regarded as having both inclusionary and exclusionary meanings.

Public places are the physical locations where we learn to live with strangers. In public places the 'diversity and complexity of persons' interests and tastes become available as a social experience' (Sennett 1986: 340). Processes of globalization such as the increase in mobility have altered relations of geographical proximity, and distance has fostered new forms of engagement between new mobile groups, hence rearranging the ethics of engagement in public places. The experience of being among strangers has been enhanced in late modern city life where processes of globalization such as travelling, migration, international business and so on erase boundaries between insiders and outsiders in cities:

> [t]his is not to say that the divisions within the city have completely disappeared but that old distributions are no longer stable and that the public spaces and sites reflect their cosmopolitan crowds. Presence and proximity is no longer an indicator of inside status, of citizenship, or of cultural membership.
>
> (Shields 1992a: 195)

The outcome of these global processes is that new configurations of public life are emerging, creating new patterns of sociability and conflict as groups lay claim on public spaces, as I discuss in later chapters.

However, I do not claim that the mere existence of co-presence and diversity does or should prompt people to engage with each other. In Sennett's (1994) view sociality is based on sensuous contact and the proximity of strangers and should not be restricted to the visual sense alone. Both Sennett (1994) and Levinas (1985) argue that sensuous proximity, or touch, fosters an ethical relationship of responsibility for the Other, as people are forced to engage with and care for each other, for example passersby tend to help someone who falls, and children are often regarded as a 'common good' which everybody feels responsible for. Yet, this is only one side of the argument, as this ideal of sensuous proximity with strangers ignores power relations that are played out through the sensuous order such as, for example, sexual harassment or racially motivated attacks. Touch in particular is often regarded as a potentially dangerous aspect which invades one's personal sphere, pollutes and demands distance and segregation. This 'fear of touching' is precisely the historical reason for the creation of ghettos in cities. It is my contention that paying attention to how social interactions are framed, supported and expressed through the prism of the senses helps us to assess the ambivalent character of social relations in public space.

The politics of representation

The third dimension that shapes public life refers to the role that public space performs as a political space for representation. As discussed earlier, the economy of access and the ethics of engagement are important features in the constitution of a potentially civil public life, which is connected to notions of democracy and citizenship.[5] Public space ideally provides the physical setting for the creation of a public sphere, which serves as a realm for political discussion, participation and representation and to which every citizen has potential access (Young 1990). This normative conception, shaped by Habermas (1974), should provide an aspatial, abstract forum of discussion amongst citizens and is best imagined as a suite of institutions and activities that mediate the relations between society and the state (Mitchell 1995). As part of the public sphere, many theorists argue, public space represents in a civil society the material location for the 'democratisation of everyday life' (Melucci 1989). As Mitchell explains:

> it is a place within which a political movement can stake out the space that allows it to be seen . . . By claiming space in public, by creating public spaces, social groups themselves become public. *Only* in public space can the homeless, for example, represent themselves as a legitimate part of the public.
>
> (1995: 115, emphasis in original)

Mitchell's argument, in short, is that public spaces are decisive in providing groups and individuals with the geographical location where they can 'make their desires and needs known, to represent themselves to others and the state – even if through struggle – as legitimate claimants to public considerations' (2003: 32–3). However, to represent oneself as a legitimate part of the public is not a straightforward process, but transforms public space into an arena of conflict and negotiation as different groups within society try to appropriate or claim their right to public space: '[a]lthough [public space] belongs to "everyone", and is historically organized by local governments, there is always great competition over its control. Whoever controls public space sets the "program" for representing society' (Zukin 1998b: 37). The control of public space can occur by the state imposing dominant meanings in space, for example through monuments or the control of spatial activities. Simultaneously, public place can be appropriated by people when inverting dominant meanings of place and reshaping them for their own purpose. In both cases this implies an alteration of the sensory organization of a place.

To make the central argument of this book clear, the sensuous set-up influences the individual's perceptions of the 'publicness' of an area. The representation of different social groups in public is based on their sensed presence. This is not only through the scopic regime but also whether their existence can be felt, tasted, smelt or heard in daily public life. Sensuous presence grants individuals a representation in public space and an active participation in shaping the public sphere, whether it is in the form of voices and languages of different groups, the historic monuments which refer to a certain social history of an area, and so forth.

The publicness of urban space is expressed and constituted through all three moments of Lefebvre's trialectic, which constantly interact; first, in the conceived space where the mental pictures, the representations of a place, inform and shape the planned, normative conceptions of how public a place should be. Second, publicness can be analysed through the perceived space – the directly observable, spatial configurations of space: the way people move and behave in an area, and the way that spatial planning allows for access, engagement and representation to develop. Yet the most comprehensive assessment of publicness arises from the lived experience of public space. Here I follow Soja (1996) who argues that perceived and conceived space misses capturing the dynamic and often contradictory spatial relations that are an important element in the production of space and which come to the fore in lived space. An examination of spatial practices and representations of space helps to understand the ideology behind particular forms of public space and how these ideologies might shape specific sensescapes and produce certain forms of spatial behaviour. Analysing lived space, on the other hand, allows us to uncover the embodied construction of complex social space and uncovers how ideologies penetrate the everyday life of a city's inhabitants. The case studies in the second part of this book are organized around these overlapping layers of space.

Having set out how to assess the public character of particular places, I now turn to the reasons for having to focus on the sensory values and practices underpinning the contemporary reconfiguration of public space in late modernity.

A new urban aesthetic

Urban restructuring is nothing new, and a number of different labels have been used in the twentieth century to describe this process, such as 'post-war urban reconstruction' in the 1950s, and 'urban development', 'urban revitalization' or 'urban renewal' in later years (see Roberts 2000 for a detailed account of the evolution of urban policy strategies). Each of these programmes has had different aesthetic values and spatial ideologies ingrained in them.

At the end of the 1980s Harvey identified a move from urban planning to urban design within urban restructuring processes. He interpreted this as a reflection of postmodern tendencies where planners moved from large-scale rational planning to an understanding of the city as fragmented and produced by multiple histories and narratives. Most importantly, urban design involves an understanding of space 'as something independent and autonomous, to be shaped according to aesthetic aims and principles which have nothing to do with an overarching social objective' (Harvey 1990: 66). Contemporary urban regeneration schemes are distinctive in their belief that design encourages and signals to outsiders economic prosperity and in their conviction that an enhanced physical environment is pivotal to the solution of so-called 'urban problems' (Hubbard 1996; Sharp *et al.* 2005; Imrie and Raco 2003; Lees 2003). Transforming the physical environment is regarded as the first and most important step in erasing the 'aura' of marginality and decline sensuously ingrained in the material texture of places:

> Physical renewal is usually a necessary if not sufficient condition for successful regeneration. In some instances it may be the main engine of regeneration. In almost all cases it is an important visible sign of commitment to change and improvement.
>
> (Jeffrey and Pounder 2000: 86)

The process of physical renewal links the spatial sensuous reorganization of place to a conscious beautification process expressed in a stylization of the urban environment. While city planning has historically been always concerned with aesthetic features, from Haussman's boulevards in Paris to Garden City developments, post-war urbanism is renowned for its emphasis on functional reconstruction over aesthetic concerns. It is only since the late 1980s that cities have witnessed 'the return of aesthetics to city planning' (Boyer 1988). This intensified aesthetic reordering of the city refers to the increased use of design, often understood as a visual script, in framing contemporary spatial identity. While city centres tend to reflect an eclectic mix of architectural styles and periods without a unifying theme, contemporary developments are increasingly characterized by an overarching and uniform aesthetic code reflected in the textures, colours, use of materials, spatio-visual landscaping and consistent theming of an area (Miles and Miles 2004).

The 'grand aesthetic coding' (Edensor 2005) of contemporary regenerated urban space is deeply interwoven with the need to produce a desirable product that can be sold on an increasingly competitive 'global urban catwalk' (see also Lees 2003; MacLeod and Ward 2002; Miles 2005). As has been widely argued in urban governance literature, cities are playing a new strategic role in the global economy as they have transformed into political actors in their own right, actively attracting, and one might argue shaping, the global circuits of capital (Harvey 1990; Castells 1996; Hall and Hubbard 1998). As the 'urban question' has moved to an emphasis on economic rejuvenation through global markets, city policies have moved away from welfare concerns and social politics to entrepreneurial policies based on neo-liberal political ideologies, often in cooperation with the private sector (Hill 1994; Hall and Hubbard 1998; Brenner 2004; Buck *et al.* 2005). The rise of the entrepreneurial city 'imbued with characteristics distinctive to business – risk-taking, inventiveness, promotion and profit motivation' (Hall and Hubbard 1998: 1–2) is directly linked to the increased competition between locations to attract foreign investment and tourism: a successful city is a globally attractive city.

Similar to the fashion catwalk, cities now contend with each other by parading made-up images of different areas of the city which advertise these spaces as favourable and appealing environments for business and leisure. Cities proudly display their new styles and designed environments on the global catwalk. But, today's hotspot can quickly be transformed into yesterday's look. The higher a city's position in the competitive structure, the more intensely its urban space will be transformed (Castells 1996; Fainstein and Judd 1999). These developments are closely linked with the rise of the symbolic economy in cities. Culture as a system for producing symbols is used as a leitmotif to appropriate spaces physically and symbolically:

> the symbolic economy features two parallel production systems that are crucial to a city's material life: the *production of space* with its synergy of capital investment and cultural meanings, and the *production of symbols*, which constructs both a currency of commercial exchange and a language of social identity. Every effort to re-arrange space in the city is also an attempt at visual re-presentation. Raising property values, which remains a goal of most urban elites, requires imposing a new point of view.
>
> (Zukin 1995: 23–4, emphasis in original)

As the culture of cities becomes the 'comparative advantage' (Zukin 1995; Kearns and Philo 1993), the architectural experience of the city becomes a further trademark to elicit consumption, so that 'eventually even city space and architectural forms become consumer items or packaged environments that support and promote the circulation of goods' (Boyer 1988: 54). It is a process that has led critics to argue that the contemporary Western city looks and feels different since the 1980s, giving rise to a distinct urban aesthetic (Hall and Hubbard 1998).

This explicit emphasis on the physical sensuous features of environments, which affect the aesthetic impact a place has on individuals, can be linked to what Featherstone (1991) has identified as the 'aesthetization of everyday life', or what Lash (1999: 125) has described as an 'architectonic' view of culture in late modernity. According to Featherstone the 'aesthetization of everyday life' is based on three developments. First, people's aesthetic sensibility has been enhanced by the increased flow of signs and images that globalization processes have produced, permeating every dimension of social life and leading to an 'aesthetic reflexivity' (Lash and Urry 1994). Hence individuals are both more concerned with and trained on issues of design and taste. Second, it refers to the aim to efface the boundary between art and everyday life, a process started by the avant-garde movements of the early twentieth century. Hence, in an aestheticized society, art can be anywhere or anything; thus functional objects, from toilets to shoes, are judged for their aesthetic value. Third, it refers to the project of turning life into art as promoted by artistic and intellectual countercultures and the construction of distinctive lifestyles. Featherstone's different aspects of aestheticization come together in the experience of contemporary cities where urban spaces act as both physical and symbolic containers and promoters of cultural life.

Lash expands these arguments by suggesting that we can no longer subjectively distance ourselves from culture, but culture is embedded through design, lifestyle and technologies in the rhythms and textures of everyday life. Culture is increasingly 'neither semiotic nor iconic but indexical, tactile and haptic' (Lash 1999: 125). Such an architectonic view of culture emphasizes the multi-sensory nature of cultural experience:

> Now culture is no longer in representations but in objects, the brands and the technologies of the information society. It lends itself much more to framing by an architectonic . . . Culture is now three-dimensional, spatial, as much as tactile as visual or textual, all around us and inhabited, lived in rather than encountered in a separate realm as representation.
>
> (Lash 1999: 149)

Aesthetics, 'the profusion of style in everyday life' (Postrel 2004), has become *the* pervasive feature of contemporary Western cities. So far, there has been a lack of research that has investigated the precise processes through which the aestheticization of everyday life is produced in specific locations. However, we first need to discuss the process of what I shall call stylization and identify what it consists of in urban areas.

On a physical level this is achieved by a redesign of the physical place where buildings or features linking a place with negative or controversial associations are made invisible. As Miles (2000) explains, this involves demolishing obsolete buildings (as has happened around the Liverpool docks), introducing new flagship projects (such as the Tate Modern at Bankside in London), readapting old buildings to new uses, such as transforming a church into a bookshop (as in El Raval, Barcelona), or transforming industrial warehouses into lofts or offices (as in Castlefield, Manchester). The physical redesign and spatial restructuring of the place can be interpreted as a 'cultural re-coding' of the place (Miles 2000). Its extreme expression is when areas of the city start to resemble theme parks (Hannigan 1998). The theming is based on a scenographic reassembly of urban components that provide an inviting backdrop to a shopping experience. 'Theming' can range from developing historical features such as the famous South Street Seaport in New York (Boyer 1992), cultural features such as the development of the Guggenheim area in Bilbao (Gomez 1998), the Disneyfication of areas such as Times Square (Sassen and Roost 1999), or the self-conscious designed environment of the Old Port of Barcelona (Degen 2008). The uniform visual landscape establishes a coherently aesthetic experiential narrative to city areas: they are 'idealised city tableaux' (Boyer 1992). Yet, as critics argue, such theming reinforces the fragmentation of the city (Boyer 1988; Sorkin 1992a). Here we have the 'disfigured city' (Boyer 1995) where fragmented elements of the city's whole are planned or redeveloped as autonomous elements of serial repetition.

Places are further restructured on a symbolic level in the form of place marketing and branding (Smyth 1994, Ward 1998). Contemporary redesign of spaces does not restrict itself to marketing the existing virtues of the city, but in fact consists of the deliberate construction of place myths which brush over 'the negative iconography of dereliction, decline and labour militancy associated with the industrial city' (Hall and Hubbard 1998: 7). Consequently, it is not only important to examine how places acquire material qualities, but also to analyse the ideological meanings embedded within their representations and place discourses: 'The production of images and of discourses is an important facet of activity that has to be analyzed as part and parcel of the reproduction and transformation of any symbolic order' (Harvey 1990: 355). Harvey emphasizes that the ways in which the city is spatialized are deeply ideological and have material consequences. The multiple dimensions and conflicts of place identities tend to be reduced to a commercialized use of culture which produces one coherent visual representation: 'As something that makes implicit values visible, however, culture is often reduced to a set of marketable images. Instrument, commodity, theme-park and fetish culture is something that *sells*, something that is *seen*' (Zukin 1995: 263, emphasis in original). However, the ways in which this image construction process might be contested at the local level are frequently overlooked, a theme that I examine in later chapters.

One of the contradictory outcomes of these urban politics is the emergence of a new geography of centrality and marginality within global cities: 'The downtowns of cities and key nodes in metropolitan areas receive massive investments in real estate and telecommunications, whereas low income workers and older suburbs are starved of resources' (Sassen 1999: 106). The social price to pay for such an unequal geography was most recently seen and felt during the 2005 riots in the suburbs of Paris and other French cities. While there is no space to examine these issues here, it is important to acknowledge that social inequalities that arise from urban restructuring processes are neither natural nor inevitable. David Harvey (1990, 2000), one of the most fervent critics of these unequal processes, argues that despite the revitalization and re-aesthetization of city areas the real underlying structural problems of places are not addressed. As I argue in the remaining part of this chapter these inequalities are not only economic but have a wider socio-cultural remit. To understand how entrepreneurial place restructuring affects the everyday life of the city, political-economical frameworks need to be expanded to take into account the role of culture in shaping social relations in the urban arena (Zukin 1995).

Commercial privatization and fortification of urban space

What are the social consequences of an intensified stylization of the urban environment for the nature of its public life? What novel forms of social life and interaction apply in these environments? In the remaining part of this chapter I discuss what I consider the most tangible manifestation of changing sensibilities in regenerated urban environments, namely the commercial privatization and fortification of public life and spaces.

The emphasis on culture as an economic motor by city governments means that there is an increased interest in developing urban lifestyles and in conceiving of strategies that promote an explicit visual consumption of public space (Wynne and O'Connor 1998; Miles and Miles 2004). In the late 1990s, one can identify an increased trend towards using images of public spaces – for example, Covent Garden, Canary Wharf, Canal Street in Manchester or Las Ramblas in Barcelona – to express the lifestyle of a particular city. These representations evoke ideal encounters portraying tourists soaking up the local atmosphere such as the music of buskers, drinking Pinot Grigio and inhaling the whiff of lavender and roses from nearby flower stalls. The sociability exercised in a place, its public life, becomes its selling advantage, the sensuously perceived 'ambience' becoming almost more important than the physical location itself. The aim is to attract an increasingly mobile public, and making this public stay and consume as much as possible through a pleasant and captivating environment. As Urry explains:

> part of the social experience involved in many tourist contexts is to be able to consume particular commodities in the company of others. Part of what people buy is in effect a particular social composition of other consumers, and this is difficult for the providers of services to ensure. It is this which creates the 'ambience' of a particular cosmopolitan city, a stylish hotel, a lively night-club and so on.
>
> (1995: 131)

Public life becomes a consumer 'good' that can be acquired by sensuously digesting the place and engaging in consumption activities. It is due to the competitive nature of the global catwalk that a 'style of life or "livability", visualized and represented in spaces of conspicuous consumption becomes an important asset that cities proudly display' (Boyer 1995: 88). This is a logical by-product of globalization processes shaping places. Yet, ironically, because of the increasing homogeneity of Western cities, globalization increases the need for places to be different while staying within the safe boundaries of familiarity. Hence face-to-face interaction and 'in situ cultural production' (Molotch 1998) – in other words, local public life and space – are crucial in providing differentiating aspects (and images) for city developers and marketers.

Consequently, public space and life need to be carefully managed. Under the heading of cultural strategies, urban space is redeveloped and given the appearance of a common public culture (Zukin 1995). The rhetoric of 'public space' hides the fact that these public places are permeated and developed by private enterprise:

> Far from being neutral guarantors of democratic legitimacy, these actors are active ideologists, mobilizing 'meaning' in both built and textual forms to provide redevelopment with an acceptable cultural alibi . . . While designers emphasize the rescue of public space in these projects, this emphasis masks the exclusion inherent in their construction of a homogeneous 'public' and the removal of land-use and design issues from public control.
>
> (Crilley 1993: 131)

One example is the way in which, in regenerated spaces, cafés, restaurants and bars are encouraged to take over pavements and squares. The consequence is that public spaces become shaped by private–commercial interests, and sensory encounters are carefully orchestrated. Participation in these spaces is singularly focused: consumption is the overall activity and aim in this kind of public life. Public space is transformed into an object of aesthetic consumption for those that can afford it, but excludes those with lower incomes. Critics are quick to highlight that these places are experienced as safe because of their homogeneous public (Zukin 1995). Trust between strangers develops out of the lack of social diversity and not out of the classic version of public life that stresses social interdependence and neighbourhood solidarity. Confidence is established by means of social control: in other words, by regulating the economy of access and ethics of engagement. An example of this is the early 1990s redevelopment of Times Square which, with funding from the Disney Corporation, has been transformed from a red light district into an 'acceptable' tourist venue featuring family retailing and global brands (Reichl 1999). While previously common meanings came from the vernacular and collective events that have shaped the history of cities, these meanings are obliterated through the commercialization of spaces (Zukin 1998b). These are places that are not conceived to dwell in but rather places to move across, consuming a constant flow of experiences. This produces 'a kind of aura-stripped hypercity, a city with billions of citizens (all who would consume) but no residents. Physicalised yet conceptual, it's the utopia of transcience, a place where everyone is just passing through' (Sorkin 1992c: 231).

A more explicit form of pacification of social diversity is the use of spectacles and open-air events, promoting collective, commercialized forms of enjoyment: 'A spatial action overcomes conflicts, at least momentarily, even though it does not resolve them; it opens a way from everyday concern to collective joy' (Lefebvre 1991: 222). Parson's (1993) analysis of the redevelopment of squares in Los Angeles shows the sinister side of this momentary standstill of conflict. The renewal of these squares was aimed at dissipating the Latino appropriation of the space. As Parson (1993) argues, Los Angeles city officials created in these squares new performance areas and food and retail operations that attracted middle-class professionals and discouraged the use of these squares by Latino people. Indeed, critical commentators claim that the range of spatial practices in regenerated environments are restricted and this is creating a setting which supports passive social contacts and passive experiences such as rest, contemplation, eating, sitting – in other words, quiet consumption (Crilley 1993; Talen 1999). In regenerated environments chance events are kept to a minimum.

Maffesoli (1996) and Shields (1992c) challenge these negative accounts, arguing that as new public spaces are emerging, new forms of sociability are fostered that need to move away from traditional conceptions. Maffesoli (1996) urges us to conceptualize differently the 'social' in late modernity and to think of people coming together as 'affectual tribes' based on aesthetic affiliations such as lifestyle rather than traditional contractual groups based on class or ethnicity. In this view contemporary life is marked by membership in a multiplicity of overlapping temporary identifications in which matters of taste and style are central. More importantly, sensuous experiences are key in these new patterns of sociability as '[t]he nature of the spectacle is to accentuate, either directly or by euphemism, the sensational, tactile dimensions of social existence' (Maffesoli 1996: 77).

In his study of the West Edmonton Mall in Canada, Shields (1989, 1992b) further stresses how sensory interactions are crucial in the appropriation of privatized and commodified places:

> The chance meeting of an acquaintance, the tactile but not too physical interaction with a crowd, the sense of presence and social centrality – of something happening beyond the closed world of oneself, motivates many who are marginal, alone or simply idle, to visit shopping centres as passive observers . . . It is the character and the 'texture' of the gathering that fascinates people.
>
> (1992b: 103)

For Shields (1992c), these spaces are 'transmodern sites', meaning that the retreat of sociality to the domestic and private, as suggested by critics such as Sennett, has never been complete, and that places such as markets, fairgrounds and nowadays shopping malls operate as important places of social engagement. Consumption is a new form of communal activity, a form of solidarity: 'In their totality, postmodern consumption sites are characterized by a new spatial form which is a synthesis of leisure and consumption activities previously held apart, performed at different times or accomplished by different people' (Shields 1992c: 6). For these writers, rather than a destruction of public life, we are witnessing a reformulation of public life that needs to be evaluated in its everyday, mundane experience.

Other urban commentators (Davis 1998; Lyon 2001; Boddy 1992) regard the reformulation of public space and life in Western cities as a more sinister development. They argue that rising concerns around fear and safety are transforming public space into an array of fortified, controlled and guarded spaces. To understand the fortification of public space, Marcuse (1997) suggests focusing on the 'boundary function' of walls rather than their physical form, as this highlights the various and ambiguous meanings and shifting purposes that walls produce: '[w]alls today represent power, but they also represent isolation; security, but at the same time fear' (Marcuse 1997: 104). Here, Marcuse draws attention to the ways in which boundaries in cities are used in contradictory ways by different social groups to stratify and shape social relationships.

One example is gated developments, or 'common interest developments' (CIDs), which have experienced an enormous increase in recent years.[6] For Rifkin (2000) gated developments reflect novel forms of social relationship which are primarily based on 'access', as 'access is becoming a measure of social relations' (2000: 115). Physically gated developments are marked by walls, fences and gates designed to restrict public access. Internally they restrict access through the strict selection processes for residents set by commercial property developers. What people do when buying into these gated communities – whether in the form of exclusive apartment blocks, retirement areas or Manchester's latest fad 'krashpads'[7] – is actually accessing a lifestyle; buying oneself into a certain type of affectual tribe which 'is a predesigned construction from beginning to end, a carefully planned commercial venture, part living space, part theatre, for those who are willing to pay an admission fee' (Rifkin 2000: 117).

Of course one could argue that gated communities are an extreme example of boundary-making. But, as Davis (1998, 1999) has shown in his influential writings on the militarization of city life in contemporary Los Angeles, hand in hand with the celebratory language of 'urban renaissance' by politicians, planners and the media, the defence of luxury has given birth to an unprecedented number of forms of security systems, surveillance and an obsession with the policing of social boundaries through architecture. Social differentiation and discrimination are played out through urban design. Here we are witnessing the official institutionalization of the 'fortress city' which is based on three features: first, an architectural privatization of physical public spaces – the control of access; second, the restraint of spontaneity and unplanned interaction in the socio-cultural public sphere – hence a control of the ethics of engagement; and, third, the reduction of the public sphere as a space for democratic representation accessible for every member in society – thus a reduction of the space for political representation. Lyon captures this development when he discusses how surveillance is being embedded in every aspect of urban life: 'on a grand scale, surveillance is steadily being globalized and, on an intimate one, surveillance now infiltrates the human body as never before' (2001: 56). That people actually comply can be explained by the metaphor of social orchestration, meaning that we are all actively involved in the mechanisms that monitor our everyday lives as '[s]urveillance is not simply coercive and controlling. It is often a matter of influence, persuasion and seduction' (Lyon 2001: 56). Power in these spaces is not so much exerted by walls, physical boundaries or coercion, but is based on seduction and constraint through cultural and symbolic strategies, the sensuous and spatial manipulation of the urban form being one of its expressions.

Conclusion

A city's public space has always been an expression of the existing political and market relations of a society. What is new in late modernity is that public spaces and the public life they support have become products, almost traded goods, on a global circuit. Public spaces are treated as marketing tools to sell a reinvented city (and its real estate). I have suggested that entrepreneurial policies have altered the urban landscape by creating 'city centres as corporate landscapes of leisure, with the re-aesthetization of the city centre accompanied by the development of consumerist "playscapes" catering for the affluent' (Hubbard 2004: 666). A similar urban aesthetic is emerging in regenerated cityscapes around the globe as the material texture of cities is redesigned and stylized according to a common script that successfully sells on the global catwalk. The consequence is that new forms of spatialization are emerging in which inclusions in and exclusions from public space operate through a diversity of aesthetic, cultural and spatial strategies that are deeply enmeshed in the sensuous-experiential geography of the city. Such developments demand a reconceptualization of public life as a spatio-sensuous encounter, and an assessment of publicness in the social production of space.

3 Sensing the city

> Sight paints a picture of life, but sound, touch, taste and smell are actually life itself.
>
> (Sullivan and Gill 1975: 181)

In the late 1980s European cities were trying to come to terms with the changes imposed by a new post-industrial economy. Many urban centres were riddled with decaying former industrial areas from Barcelona's El Raval, to London's Hoxton or Manchester's Castlefield. Imagine for a moment that you are walking through one of these neighbourhoods. You are almost certainly a lone tourist; the reputation is such that few outsiders dare enter. This is a no-go area, a desolate place; its streets are narrow, littered with rubbish and overgrown with weeds. You are in Castlefield, walking through the empty shells of ruined buildings covered with grime. It is a barren space whose residents left a long time ago when the wharfs stopped functioning, when the factories stopped producing. You are in El Raval, avoiding prolonged eye contact with those on the margins of society: the poor, the old, the prostitutes and the drug addicts. Breathing in the stale air, you wander through a scattering of family-run grocery shops and work-sheds, outnumbered now by the barricaded doors of businesses that have been unable to make ends meet. These are Sassen's (1991, 1998) geographies of marginality, places of persistent public and private disinvestment.

Now fast-forward five years to the same neighbourhoods. Their environment has been radically transformed. Construction noise fills the air, old buildings have received a new coat of paint, some have been demolished and replaced by sparkling new ones. Art galleries, boutiques, cafés and restaurants have replaced factories, workshops and empty shop facades. New residents have moved into El Raval: immigrants, young people and those working in the creative industries, as new media offices, design studios and architecture practices have settled in the old workshops. In Castlefield dilapidated wharfs have been converted into luxury city residences or designer bars. Eva, an art gallery worker, describes the changes in El Raval:

> In the last few years they have been trying to give a different meaning to this place. They are trying to get people into this barrio. You've already seen all the construction work everywhere. They are fixing the whole barrio to give it more light; more life. They are providing new facilities so that people now want to come to this barrio. What I mean is that from a no-go area this is becoming one of these *barrios* fashionable to live in.
>
> (Eva, art gallery worker)

Perceptions of places are mapped through sensory experiences. Yet, despite the fact that 'how a city looks and how its spaces are organized forms a material base upon which a range of possible *sensations* and social practices can be thought about, evaluated and achieved' (Harvey 1990: 66, emphasis added), literature on urban redevelopment has largely ignored the importance of sensory dimensions in defining urban experience (exceptions are Law 2001, 2005).

While the purpose of the last chapter was to determine *what* defines a space as public, the aim of this one is to address *how* to examine the changing aesthetics of contemporary public places. The chapter starts by suggesting a reworked notion of aesthetics that emphasizes and develops the sensory dimensions of the term. By reviewing the legacy of earlier approaches to the senses and urban experience I argue that to analyse the city we have to move away from a fixation on and preference for the visual, to the analysis of a more comprehensive multi-sensuous experience – what I describe as a 'socially embedded aesthetics'. I follow this by showing how each individual sense mediates a unique form of experience and thereby produces a distinct urban sensescape. These arguments are expanded in the final sections of the chapter by suggesting tools such as 'place gestures' and 'rhythms' to show how senses map the socio-cultural spatialization of places.

Towards a 'socially embedded aesthetics'

A pervasive feature of contemporary urban redevelopment is the excessive importance given to the visual. This is due to postmodern trends in urban planning that have led to an increased stylization and theming of the urban landscape and have been summarized as an intensified 'aestheticization' (Boyer 1992; Featherstone 1991; Jameson 1984; Zukin 1995). Critiques of these developments highlight three main features: first, the reduction of the multi-sensuous city to the visual sense alone, as suggested by Zukin (1995: 10):

> Visual display matters in American and European cities today, because the identities of places are established by sites of delectation. The sensual display of fruit at an urban farmers' market or gourmet food store puts a neighbourhood 'on the map' of visual delights and reclaims

it for gentrification. A sidewalk café takes back the street from casual workers and homeless people.

Zukin (1995) illustrates how the identities of contemporary public places are redesigned and transformed through the power of vision. She describes how these visual transformations and linked activities displace other experiences and spatial practices. The city is conceived here foremost by the visual sense, deeply disconnected from other senses. Sennett (1990) describes this as the 'compulsive neutralization' of modern city planning. The aim of this city, symbolized by the increased use of glass, is to achieve complete visibility, without exposure to other senses. Glass not only detaches those inside the building from the immediate surroundings but also 'seeing what you cannot hear, touch, or feel increases the sense that what is inside, is inaccessible' (Sennett 1990: 110). The neutral city is based on an encompassing and homogeneous visuality that reduces the complexity of the city into an easy, clear and digestible vision. Expressive differences between particular localities are subsumed under a single visual whole, often the vision of planners. Sensuous difference is neutralized through visual uniformity.

The second feature that critics denounce is a manufactured visuality in the proliferation of 'spectacles' to sell the contemporary city (Harvey 1990; Zukin 1995; Hannigan 1998). Here we encounter a deliberate visual staging of events disengaged from any political meaning, merely based on entertainment, and pacifying any critical potential through visual pleasure and seduction: 'The spectacle . . . is a visual delight intended to immobilize our attention in the act of "just looking"' (Boyer 1992: 192).

Last, critics denounce the fact that vision has been the most technologized sense in modern society. Virilio, for example, criticizes the development of 'sightless vision' in social life in general, and in cities in particular, where perception is automatized by 'delegating the analysis of objective reality to a machine' (1997: 389). The city is transformed into a visual-technological topology where all non-visual senses are denigrated and subsumed under the visual: 'tact and contact give way to televisual impact' (1997: 386), meaning that the visual sense becomes crucial in gauging impressions. For Virilio, this results in a vision that has become independent of human agency.

It is important to highlight that these discussions on visuality and the city do not critique the visual sense per se, but rather denounce the hegemony of a specific sort of visuality in defining late modern urban experience. Humans and senses are subjected to a dominating visual sense which controls through surveillance and, most importantly, controls by annihilating a diversity of experiences and meanings that could emerge from the other senses. While not disagreeing with these claims, I believe that these critiques are in danger of falling into the same trap they are criticizing: giving precedence to the visual in their analysis of urban change. As Howes (2005a) maintains, while visual display might have been the primary source of consumer culture in the past, Western culture is experiencing a commercialization of all the senses. Let me be clear: the aesthetization of the urban landscape encompasses as much a visual landscaping as a conscious orchestration of particular sounds, smells, tastes and feelings in the urban environment. However, writings on the aesthetization of the urban landscape tend to assume an imposition of hegemonic visual meanings onto a relatively passive audience. Yet little empirical research has been conducted on what the 'aesthetic effect' consists of, and how people use and experience these environments (Degen *et al.* 2008).

To address this neglected issue I suggest a return to the original Greek meaning of aesthetics: 'the perception of the external world by the senses'. For the remainder of this chapter, I shall reframe 'aesthetization' within this original meaning and develop the theoretical tool of 'socially embedded aesthetics'. This concept captures two features of the sensuous analysis of urban environments: the embeddedness of our daily experience in a sensuous world and the social character of the senses – the ways in which senses shape and are shaped by social relations. This will provide me with a base to illustrate how physical transformations in regeneration schemes, hence the reorganization of sensuous geographies, are an important way in which social groups and place identities get shifted around and controlled. In what follows I briefly outline most influential academic perspectives that have engaged with sensuous experience, and review those aspects that are relevant for the development of a 'socially embedded aesthetics'.

Two of the most significant writers to link a sociological analysis of the senses and urban experience have been Georg Simmel and Walter Benjamin. Their writings on the city in the early twentieth century emphasize the constant sensory engagements that structure the individual's experience of modern urban life. According to Benjamin, 'it was through the jostling crowds of the city, and the decaying fabric of its buildings as they past into obsolescence that one could understand modernity' (Leach 1997: 24). The sentient body in the modern city is constantly challenged by unexpected experiences, by ephemeral sensuous arousals, when walking through the anonymous metropolitan crowd. Indeed, the rise of the modern city was defined by technical innovations that dramatically transformed urban perception. Benjamin cites the example of the safety match, or the camera, where one movement triggers off a wholly different sensory experience which alters our encounter with the world. The camera for example, captures a fleeting moment in time, while the match provides warmth and light in a second. Haptic reactions are here clearly connected with changes in visual experiences, such as the increase of neon advertisements, or the traffic in cities which makes the human being walk, machine-like, 'in a series of shocks and collisions. At dangerous intersections, nervous impulses flow through him in rapid succession, like the energy from a battery' (Benjamin 1997: 31). Thus, for Benjamin, the emergence of the modern city has transformed the practico-sensory body in that 'technology has subjected the human sensorium to a complex kind of training' (Benjamin 1997: 31).

Simmel's early work, first published very early in the twentieth century, similarly analyses how the senses are involved in mediating and ordering human interaction (1997). We get involved in interactions because we have sensory effects on each other as 'every sense delivers contributions characteristic of its individual nature to the construction of sociated existence' (Simmel 1997: 110), a point to be examined later. Of particular interest for the argument here is his discussion of the experience of the modern city in terms of its constant sensory stimulations. In the essay *The Metropolis and Mental Life* (1971b) Simmel famously describes how modern city experience is unique in its sensory overstimulation. The great amount of sensuous stimuli that the individual is confronted with leads to 'the intensification of emotional life due to the swift and continuous shift of external and internal stimuli' (1971b: 325). The individual protects herself from an overstimulation by adopting a blasé attitude or aloofness to the environment. It is not the time or place to judge this description, but what Simmel points out is that sensuous experience shapes social interactions and that to understand city life one needs to investigate 'the relationship which such a social structure promotes between the individual aspects of life and those which transcend the existence of single individuals' (1971b: 325). The important legacy that Simmel and Benjamin provide is that they draw connections between the experience aroused by the stimulation of the senses and broader social changes, hence highlighting the intrinsic social character of the senses. Not only do they illustrate that the senses play a crucial part in daily experience but, more importantly, they argue that urban public places are the media through which we can experience and assess changes in society.

A few years later, phenomenology – the study of how phenomena, or the world itself appears – provided a further step towards an understanding of the 'corporeality' of space by arguing that space, rather than being an absolute, neutral backdrop to social life, is actively produced through lived experience. This approach countered the tendency to perceive space as an abstract cognitive construct remote from bodily sensations (Merleau-Ponty 1969). Following Kant, this philosophical school believed that inner perception was impossible without outer perception. Its aim was to work out a language to describe the world as we perceive it through the senses, without distorting this direct experience by reflection – explanations or analysis. For phenomenology a description of sensory experience is an essential feature of understanding reality. We have to imagine the perceived thing as a 'gestalt': a mental pattern, that is constituted by the interplay of different sensed aspects 'and each does something in its own place and moment to contribute to the composition of the thing' (Lingis 1994: 4). Merleau-Ponty talks therefore about 'the sensible essence of things': the way in which each element of a perceived thing adds to the unique being and meaning of the object. If we eat an orange, its colour, its porous skin and its fleshy liquid inside represent each of the sensory features that make up an orange and help us to differentiate it from an apple.

For phenomenologists the body and the world are in a reciprocal relationship. However, bodies are not randomly aroused by stimulation but organize their perception actively (Lingis 1994). This means that sensing is an action in which the body exerts a crucial role actively making sense of the world, as Lingis illustrates with the sense of touch:

> The tactile datum is not given to a passive surface; the smooth and the rough, the sleek and the sticky, the hard and the vaporous are given to movements of the hand that applied itself to them with a certain pressure, pacing, periodicity, across to certain extensions, and they are not patterned ways in which movement is modulated.
>
> (Lingis 1994: 8)

This is an important feature of phenomenology, as it points out that people are active in the creation of meanings and that an essential part of being human is linked to the perception of our surroundings, the experience of place (Pile 1996). Space in this conception is not an external force that inscribes the body but is 'a lived and shared dwelling whose "effects" cannot be understood or accounted for independently of the human action which animates them' (Crossley 1996: 107).

A drawback of the phenomenological approach is the strong emphasis that is placed on describing the material, existing, real world rather than the use of interpretation. In its attempt to describe perception as it is, it does not question the ideological construction or social values that are inscribed in the environment and therefore does not offer room for a critical stance (Derrida 1973). It also refuses to comment on imagined, not immediately existing spaces, such as daydreams, or more importantly past experiences or memories.

So what does the phenomenological approach offer for the development of a 'socially embedded aesthetics'? First, the importance it ascribes to description as a valid form of inquiry, yet an often neglected method in urban analysis: 'We should start with *description*. I am permanently surprised at how little good description there is of the galactic metropolis in the academic literature' (Lewis cited in Knox 1991: 204, emphasis in original). Second, it draws attention to experience as a corporeal process and emphasizes how different senses work actively together to develop a meaningful understanding of the environment. Hence, by regarding consciousness as 'embodied', phenomenology draws attention to the expressive role of the body, linking thereby consciousness, experience and perception. Finally, phenomenology highlights that the body is dependent on cultural repertoires and skills but is equally responsible for their reproduction (Crossley 1996).

Phenomenology focuses on the individual body and experience. Yet, place experiences are linked to cognitive processes, the associations and knowledge that we connect to particular places: in other words, the stories, myths and reputations associated with a particular neighbourhood. Humans tend to map their environment with mental processes, often defined as 'images', by which we 'assemble, understand, remember and use spatial environmental information' (Pile 1996: 27). More developed approaches quickly recognized that cognitive mapping is not merely a subjective activity but meanings are shared by groups of people as these cognitive structures are expressed in language. Lynch summarizes:

> Cognition is an individual process, but its concepts are social creations. We learn to see as we communicate with other people. The most interesting unit to study for environmental cognition may therefore be small, intimate, social groups who are learning to see together, exchanging their feelings, values, categories, memories, hopes, and observations, as they go about their everyday affairs.
>
> (Lynch 1976: v)

Groups learn to see together and the communicative exchange in a group informs the feelings and identity given to a place. Thus representations of places are intersubjectively constructed through the exchange of narratives, which produces a spatial social imaginary. Different social groups will have different spatial social imaginaries depending on their relationship to places, which can clash in the restructuring and recoding of places. This is a theme I discuss in the later part of this book.

These diverse approaches highlight the significant role played by the senses in our active engagement with and experience of the environment, as well as illustrating that these sensuous experiences are socially framed. However, they do not provide an insight into the ways that the senses organize or stratify urban spaces socially and culturally.

The city of senses

I would now like to invite the reader on a sensuous journey. In order to open up more expansive ways of thinking about the relationship that each sense constructs between the self and the city I propose to imagine the city of touch, of sound, of smell, of taste and finally the city of sight. The purpose is threefold. First, it draws attention to urban space being constituted through the sensory-practico body. While there has been an expanding literature on embodiment, the body's senses have largely been ignored or 'assumed to be an intrinsic property of the body – a natural and unmediated aspect of human being' (Law 2001: 266).

Second, I discuss the particular spatial experience that each sense offers and its individual importance in framing the characterization of place.

Senses are central ingredients in the organization of quotidian experience. Yet inherent in the senses is a latent ambiguity, in that their meanings and experiences are constantly open to change, and a caressing touch can quickly turn into a sexual threat. It is important to emphasize at this point that bodies of course differ greatly, and that age, dis/ability, ethnicity and gender deeply influence how bodies sense and are sensed, which leads to differentiated sensuous spatializations of places.[1]

Finally the analysis of the relationship each sense establishes with an urban environment aims to highlight the sensuous encounter in cities in terms of 'sensescapes', a term shaped by Porteous who argued that, similar to the term landscape, 'smellscapes suggest that, like visual impressions, smells might be spatially ordered or place related' (1985: 359). The notion of sensescapes depicts the layering, superimposition and simultaneous presence of several sensuous experiences in geographical encounters.

I deliberately leave the city of vision to the end of my discussion because Western society has had an ambivalent relationship to vision. Vision has historically been regarded as the most important sense, and has had a tendency to be placed at the top of a 'sensorial order'. In the Aristotelian hierarchy of the senses, sight and hearing were considered the 'human senses' – the noble senses – whereas smell, touch and taste were regarded as inferior and animalistic (Synnott 1991). Although such a strict distinction is not the case anymore, or at least not spelt out in such an overt way, claims based on visual observation are still the predominant mode of inquiry in the sciences. Most of the time fieldwork is still based on seeing places, with the implication that the manifold qualities of place are reduced to the visual alone (Ingham *et al.* 1999; Porteous 1985). The non-visual dimensions of experience tend to be ignored or relegated to the world of literature, to evoke the 'character' or 'atmosphere' of places. By inverting this 'sensuous order', and by leaving the city of vision to the end of my account, I hope to sensitize the reader to the fact that in our everyday perception most of us 'see' aided by the interplay of all the senses.

The *city of touch* is based on (dis)connections. The tactile sense involves a twofold process, since at the same time that we touch we are touched. It is the most primordial sense with which we establish a relationship with what is outside us through the whole of our body: 'Hands and feet, lips and face; the skin mediates touch inside and out, environment and individual in continuous kinesthetic flow' (Foster 1998: 55). Through our sense of touch the surrounding environment becomes alive and a textured part of an overall bodily experience. We can feel the cold aura that concrete exhales, benches become surfaces too hard to sit upon, the paved ground resists our every step. Yet touch can also be non-directed. We bump into things and are startled by their materiality. The haptic sense makes us aware of the feeling of place as it informs us about the temperature and surfaces surrounding us. Thus the world around us communicates with its material (in)consistency and evokes certain experiences and sentiments to us; touch informs us about 'being' in the world (Rodaway 1994). Touch is the 'sensorial snail' (Classen 1998), as it involves feeling one's way gradually through an object, rather than capturing it all at once. The sense of touch permits gradual revelation and pleasure or disgust in its discovery.

Touch has been described as the most reciprocal sense, and is perceived as an important element in developing democratic public places. Physical contact, or closeness, plays an important role in interpersonal and environmental relationships. Our relations to other city dwellers are established through a careful monitoring of distance and proximity. The sense of distance varies between different cultures that have different demarcations of interpersonal space (see Hall 1969; Rodaway 1994 for a detailed account). One could argue that tactility is needed to establish a close relationship with other people; it is the most intimate sense, since we touch to express caring and love. Yet this intimacy can quickly turn into danger and fear when it is perceived negatively and threatening. Touch necessitates movement – we move towards, or something approaches us to establish contact; a certain form of transgression has to occur. With the sense of touch, as with taste, inside and outside become one, they impregnate each other, and contamination becomes a felt danger. Fear of touch plays an important role in the spatial division of the city, as Sennett (1990) has discussed in regards to the Jewish ghetto in Venice. During Renaissance times Venetians believed that the Jewish body was polluting Christian values and community. As I discuss in the next chapter, they isolated the contagious bodies by enclosing the Jewish population into a ghetto, which poignantly illustrates how urban environments are ordered through features of tactile proximity and distance.

The *city of sound* surrounds us; we cannot shut out sound like sight; 'the pulse of a city' is constantly with us. While vision makes us aware of our distance to an object, sounds involve us in the world. Sound has a more emotional dimension than the other senses. It transforms the space around us from inside us; we are moved by music, by a crying baby. The aural world is an eventful world (Ong 1971). Things need to move, vibrate, in order to emit a noise. Sounds imply action, and in a city, footsteps, the opening and closing of doors, voices, a car, and so on generate a sonic presence of people (Borden 2000) that can indicate safety or danger.

Sound is not just sensation: it is information. We do not merely hear, we actively listen (Rodaway 1994). However, in the auditory sensescapes we are often confronted by a variety of different sounds, which cannot be individually distinguished and are experienced as a background hum. Hearing is the most dynamic of the senses as it can detect change in both intensity and tone. Described as the most democratic of senses as it is supra-individual, it connects people (Featherstone and Frisby 1997). Similar to smell, sound cannot be controlled: 'It is too changeable, too transient to be dominated – as one dominates landscape through sight – [sound] can only be attended to and engaged with' (Classen 1998: 142). Thus, it is a subversive sensescape:

> [as it is] flowing outside the bounded visual system imposed by architecture, causing people to do things and to create possibilities. The effect, particularly as the aural city often changes more rapidly than the visual city, is to impose a different rhythm on to the visual and physical rhythm of the city, leading in turn to layering, interweaving and discontinuity, as when sound transgresses barriers.
>
> (Borden 2000: 30)

In the urban auditory world a number of noises overlap; certain sounds dominate others because of their loudness or repetitive actions, quieter sounds disappcar in our experience and thereby hide a part of the environment. Space can be demarcated by sound and intensified, as for example with a street festival where the street obtains a particular identity through the diverse 'audioscapes' caused by the event. Sound, or its absence, can divide or link two separate spaces: inside a house the noise of outside traffic, such as police sirens and beeping cars, disappears or filters through, and questions what is inside or outside, private or public (Borden 2000). Last, the transformations of a city can be read through its changing sonorities. The mechanical rattling of cranes, for example, signifies action and change, whereas the development of new urban control systems can be read from the soundscapes of the city where the whistle and shouting of a policeman have been replaced by the silent computer systems of traffic control (Fortuna 2001).

The *city of smell*, similarly to the city of touch, presupposes a certain intimacy between subject and object: a relation of reciprocity. Smells are often associated with the notion of transition; olfaction represents the crossing of a threshold (Howes 1991). It is at the threshold of a room that one most notices its odour; after a few moments inside the scent disappears as one adapts to it. Most of the time it is the outsider that becomes aware of the smellscapes of a place: the guest who walks into a room, the tourist who identifies the particular odour of an area. Yet smell is the most evasive sense to describe. It is a feature that always escapes; it is formless, it cannot be articulated, it cannot be defined in static terms. There are no distinctive smell adjectives but only 'smells like', '[i]f we say, "it smells sour" then this only means that it smells the way something smells which tastes sour' (Simmel cited in Featherstone and Frisby 1997: 118). As the quote highlights, the sense of smell is strongly linked to taste, emphasizing the relational character of senses.

With the sense of smell and hearing we are exposed to stimulation from outside and we cannot refuse it. Smells and sounds, while being produced by an object, escape the object, and are not prisoners of an enclosed form (Howes 1991). Odours are often described as the most pervasive sensory stimulations, as they do not respect boundaries and are not governable. Smells permeate our personal realm as they infiltrate our body, our clothes, at the same time as they are fragmentary, non-continuous experiences, episodic in time (Porteous 1985). While they penetrate our consciousness, we cannot pin them down and control them as we can control our sight or the food we eat or objects we touch. Odours are immaterial; they merge, can linger over time or temporarily vanish. The sense of smell defies visuality and tactility in that you do not need to be near an object to surrender to its odour. Smells cross boundaries, are always 'out of place', always on the move: '[S]mells are thus ideally suited to expressing the notion of contagion or action at a distance. And the reason for this . . . is that they are always "out of place", forever emerging from things, that is, crossing boundaries' (Howes 1991: 140). Odours cannot be stopped from entering a space – smells pervade place. Consequently, as Bauman reminds us, odour in cities became the sinister Other of everything modernity stood for, the opposite of control and order: 'Smells share with Simmel's strangers the upsetting habit of coming unannounced, outstaying their welcome, arriving now and refusing to go away later' (1993: 24). As I discuss in the next chapter, this is why smellscapes are often used to support class distinctions and to order the city. Visual distance prevents contamination; not so odour, which defies distance and travels as quickly as the wind direction changes. When examining the history of odour (see Corbin 1986; Lowe 1982) we can see that natural odours, i.e. those smells that we cannot restrain such as body odours, smells of excrement and so on, signified barbarity, whereas artificially created odours and those applied in a controlled manner, such as perfume, signified civilization.

Taking a more positive stance one could argue that odours individualize and identify objects and particular places, and take control of them. They are linked with our emotional states, as olfactory receptors are linked to the brain's limbic system (Porteous 1985). Thus smell will reinforce the particular character of and our emotion towards a place. Scents trigger memories, familiarity, attraction or dislike to people and objects. On a subjective level smells cause changes in the way we experience a place (Howes 1991). Our nose will also help to ignite different experiences of place, and spark off an olfactory memory which 'bypasses the conceptual part of the brain, the neocortex, whereas sight and sound do not. The result is that odour related memory is more immediate, and called up more directly than memories connected to sight and sound' (Brady 2001: 19). Smells provide us with a more fluid experience of space and time. They subvert the immediate experience of place by making individuals relate to other places and times.

Related to the city of smell is the *city of taste*. Not only can we actually perceive a certain urban flavour, for example when we enter London we might perceive the taste of smog, or in a Mediterranean city we are able to taste the salt in the air (amidst the smog). More importantly we savour the city by eating and drinking our way through it. As we digest food or feel the taste of a smell in our mouth, place and person become one. Yet to taste is the most egoistic of all senses as: 'what the individual eats, no one else can eat under any circumstance' (Simmel 1994: 346). To taste, one needs to physically posses an object so that it becomes part of one's personal realm.

The sense of taste becomes acquired over time. We grow up getting used to specific tastescapes which we link emotionally to our sense of being and place. This might be the reason why tastescapes represent most overtly sensory cultural politics, as Law (2001) illustrates in a article on sensuous practices by Filipino domestic workers in Hong Kong. Filipino food becomes a contentious subject in Chinese homes where many domestic workers are forbidden to cook this food. Therefore 'it articulates national identity in a way that it could not in the Philippines' (2001: 278), and the consumption of Filipino food in public places 'becomes a positive signification of cultural difference' (2001: 276).

Tastescapes play an increasingly major role in the sensuous landscape of regenerated neighbourhoods. They provide the possibility of experiencing 'Otherness', such as in the traditional Chinatown or Little Italy areas of cities. Since the 1990s, with the rise of café-bar culture, tastescapes can signify an economically successful urban life. Essential elements in the provision of a new urban global aesthetic are public places surrounded by cafés where one can sit outside and observe the surrounding daily life while titillating one's tastebuds. The identity of place is increasingly portrayed through its gastronomic reputation: 'Restaurants have become incubators of innovation in urban culture. They feed the symbolic culture – socially, materially, and spiritually' (Zukin 1995: 182).

Finally, in the *city of vision*, sight situates the viewer outside what he sees and demarcates a physical distance. Similar to the other senses, sight is a subjective sense (Featherstone and Frisby 1997). Seeing is a selective process as we consciously direct our vision. The sense of sight promotes a fast and immediate appropriation of a surrounding environment or object which only lasts as long as the object is within our field of vision. The implications of these features are that the visual format gives an impression of transparency – it establishes what is tangibly present and absent – thus the illusion that things are exactly as they look (Stewart 1995). With vision being the most immediate sense, it is the sense that makes most tangible the passing of time. While a specific smell or an old tune can remind us of the past, we perceive the passing of time through vision:

> what we see of a person is the lasting part of them; as in a section through geological strata the history of their life and what it is based upon as the timeless dowry of nature is revealed in their face. The variations of facial expression do not correspond to the diversity of that which we ascertain with our ears. What we hear is a person's momentary character, the flow of their nature.
>
> (Simmel 1997: 115)

Paradoxically, while vision is important in establishing presence in the immediate now, it proves difficult for our memory to recollect concrete mental pictures. The reason may be that visual geographies are dynamic in nature, as we constantly move the eye around and the individual thereby composes and selects an ephemeral view in the present moment which is difficult to recollect once that moment has passed.

The ocular sense provides pleasure as we appreciate different forms, colours and textures of the environment through the eye, letting our gaze linger on an object we like or shutting out what we want to ignore or regard as an unpleasant sight. Sight is further understood as a mutually enriching encounter that smoothes out and democratizes power relations, as people interact by looking at each other – what Jay (1993) describes as 'dialogic' specularity. For Simmel, vision is an important component in facilitating everyday sociability in public spaces, as when people's eyes meet this produces the 'most complete reciprocity', and as this sensuous interaction momentarily engages two or more people:

> The extremely lively interaction, however, into which the look from one eye to another weaves people together, does not crystallize in any objective structure, but rather the unity that it creates between them remains directly suspended in the event and in the function.
>
> (Simmel cited in Featherstone and Frisby 1997: 111)

Vision, of course, can also act as the least reciprocal sense and signifies non-engagement or refers to unspoken power relations when we evade eye contact.

Sight has played a major role in representations of urban place and in its commodification. Visual technologies have been developed that capture and situate place experience. Maps, photographs, film and so on are regarded as reflections of the visual character of a place under a false aura of rationality and objectivity (for a critique see Berger 1972; Jay 1993). However, images are not neutral. They select a slice of reality, are enhanced and reworked, and serve to manipulate our emotions as film or advertising shows (mostly supported by sound). This is opposed to the ways in which we gradually appropriate places and objects by moving through the street, by getting involved in its visual deciphering with the help of the other senses. Thus, the represented experience of a visually pleasing square on a photograph can be contradicted in the lived experience by the surrounding traffic noise and smells coming from nearby rubbish bins. To emphasize again my point: our senses work together to contextualize our experience of place.

Place gestures

Our geographical experience and imagination are based on the interplay of body, senses and place. When experiencing the city our bodies are confronted with a plurality of sensuous engagements, from the beeping of cars to a perfume passing by or tactile experiences such as cold wind, or the pressure of crowds. So how are we to assess and analyse these 'sites of embodied intensity' (Allen 1999)? These perceptions are not unique to the individual, or merely subjective, but are shared by all of us as we experience these sensory stimuli through a common medium, the body (Nast and Pile 1998a).[2] In fact, following Rodaway, we can say the body acts as a mediator between the self and the world:

> The body contributes both to spatial and temporal perception, being like a ship and its anchor in our lifelong geographical experience. It mediates between us and the environment, giving us access to a world beyond itself. In fact, without our bodies we would have no geography – orientation, measure, locomotion, coherence.
>
> (Rodaway 1994: 31)

The combination of different senses contributes to our spatial orientation, an awareness of spatial relationships and the appreciation of the qualities of particular places. This leads Rodaway to suggest the notion of a 'sensuous geography' to describe 'an interaction with the environment both as given to the senses and as interpreted by the senses themselves in conjunction with the mind' (Rodaway 1994: 26). Yet senses never work on their own but are framed in context and in relation to a reference – objects which they define.

Urry (2000) used the notion of 'affordances' of environments or objects to draw attention to the fact that senses connect hybrid objects, the human and non-human. The term, first developed by the ecological psychologist James Gibson (1986), suggests that the composition and layout of environments and objects 'affords' certain types of behaviour. Subsequently, there is not simply an objective reality 'out there' but instead affordances are qualities in the environment perceived relative to the observer. These affordances are not inscribed in space, but rather activated through people's sensory experiences, by moving through, touching, smelling, tasting, hearing and seeing objects and places. This supports phenomenologists' claims that individuals are involved in *active sense-making*. It is the senses that connect human capacities with objects. Examples of affordances are: a large open square that fosters the gathering of large groups of people; a tinted glass building that allows viewing of the outside but not the inside; or a particular building that affords and triggers certain memories. The affordances in places are never fixed but constantly reorganized, both by the physical set-up of the place and by the way different users either resist or submit themselves to these affordances. As I discuss later, the reconfiguring of public space involves a reconfiguration of affordances and resistances which produce new sensescapes.

Consequently, space can be viewed as performative in that it affords certain practices and sensory experiences, but at the same time it is also performed through the actions and experiences sensed by the bodies that participate in this space. Grosz (1998) describes this relationship as an interface between body and city, a constant filtering between one and the other 'whereby a body sensually and knowledgeably ingests the city as the city, in another sense, ingests it' (Nast and Pile 1998b: 6). Being in a public place is an intercorporeal encounter. It is the sensuous fabric of the city, in terms of both the relationship established between objects and the walker's 'floating attention to events that constantly take place in the street' (Mayol 1998: 13), which provides the capacity to move the individual emotionally and transports the walker to other places and times. It is not in the abstract image of the city, as conceived by planners and politicians, but in the daily arrangement of urban components that can be sensed that the city becomes a place filled with the constant dialectic of revelation and concealment.

These kinds of non-quantifiable experiences of place are often described as the atmosphere or mood that places and settings evoke. It is a relational moment created by objects, people, discourses and practices – always transmitted first by the senses and thus first understood and then interpreted (Albertsen 2000). I suggest that to examine the creation of atmosphere in places it is fruitful to think of the affordances created between places, buildings and people as *place gestures*. The expressive properties that the sensory experience of place provides become an inferred meaning: that which cannot be directly said or even explained. Gestures are affective suggestions in space, moments of evocation that are materially performed. They are playful innuendoes that evoke a certain reaction from the person that has noticed them. They are sudden moves that can be there one second yet rapidly disappear, never to be recaptured, never to be fixed. It is in the immediate lived experience of public place that we engage with these multiple time frames simultaneously. The spatial order of places gets constantly subverted by time: 'places are never simply confined to the present time and immediate location, but are charged with the (collective) memories and associations of other times and places. Sensory perception opens onto a disjointed proximity in space-time' (Ingham *et al.*1999: 287). Both past memories and potential futures can be activated through a sensory experience – a noise, an unexpected view, the touch of a wall – and can subvert the immediacy of the present moment, of our present being. Gestures can be haunting presences of either past times or future promises, always influencing the affective perception of place, almost haunting the present moment: 'Being haunted draws us affectively, sometimes against our will and always a bit magically, into the structure of feeling we come to experience, not as cold knowledge, but as transformative recognition' (Gordon 1997: 8). By using the notion of 'place gestures' I draw attention to social life not only being constituted by the empirical quantifiable but also by the absent and suggested (Degen and Hetherington 2001). It would be wrong, as Gordon (1997) suggests, to relegate these features in social life only to the subjective realm, as, for example, power relations are mostly exercised covertly and work through (un)graspable material relations in everyday life.

Similar to human gestures whose meaning we can intepret, our experience of place gestures depends on our affective relationship with the place and the familiarity we have with an environment. Senses engage us with traces of time inscribed in this urban landscape that are already vanishing or yet to come. Analysing gestures of buildings or places thus allows us to capture both the multiple and overlapping sensescapes and the time-embedded socio-cultural meanings experienced by different user groups. It provides us with a device to show how hidden cultural-aesthetic power structures are performed in and by the sensuous landscape. Hence the promise of an engagement with the senses is not simply one of disclosing the place related qualities of the sensory world, but also of revealing the role of different sense relationships in creating and defining particular places at particular times.

Methodological discussion: rhythmanalysis

Lefebvre's method of rhythmanalysis (1991, 1996) offers us some clues as to how to analyse the elusive processes that inform the socially embedded aesthetics of places. At the end of his lifetime Lefebvre increasingly focused on the senses, especially in his development of the concept of rhythmanalysis with which he sought to examine the spatio-temporal relationships between the body and space. His premise was that social space is experienced (and constituted) first of all through the body:

> The whole of (social) space proceeds from the body, even though it so metamorphoses the body that it may forget it altogether – even though it may separate itself so radically from the body as to kill it. The genesis of a far away order can be accounted for only on the basis of the order that is nearest to us, namely the order of the body. Within the body itself, spatially considered, the successive levels constituted by the senses (from the sense of smell to sight, treated as different within a differentiated field) prefigure the layers of social space and their interconnections. The passive body (the senses) and the active body (labour) converge in space. The analysis of rhythms must serve the necessary and inevitable restoration of the total body.
>
> (Lefebvre 1991: 405)

Thus for Lefebvre space is a product of the human body 'as a perception *and* a conception, not simply as the physical imposition of a concept, or a space, *upon* the body' (Stewart 1995: 610, emphasis in original). Rhythm-analysis is an intrinsic part of exposing the actual production of space, as Lefebvre seeks to capture the dynamism of environmental sensing and the embeddedness of social relations in the senses and in space.

The rhythmanalyst does not restrict his/her observations to the visual, but listens and perceives the movements in everyday life, the cyclical comings and goings of people, of nature, the subtle transformations of space. One observes, perceives, to attain a particular state of awareness (Allen 1999), to capture the gestures of place. There are two processes that rhythmanalysis pays attention to. First, rhythm in terms of activity, and second, rhythm in terms of how the senses map a particular landscape. Activity rhythms refer to the daily movements and the everyday, repetitive spatial practices of people: the comings and goings of people to and from work, the rush-hour traffic, lunchtime, the garbage men collecting rubbish, and so on. Jane Jacobs famously describes this as a 'sidewalk ballet':

> an intricate ballet in which the individual dancers and ensembles all have distinctive parts which miraculously reinforce each other and compose an orderly whole. The ballet of the city sidewalk never repeats itself from place to place, and in any one place is always replete with new improvisations.
>
> (Jacobs 1961: 60–1)

The mix of superimposed, parallel flows that are repeated each day confer the place with a specific rhythm and give those who live, visit and work in a place a sense of location. It is this repetitiveness that produces spatial rhythms. What interests us especially about the fluctuations of rhythms is that once the rhythmanalyst has established the interaction of rhythms, the next step is to determine what kind of relationship these rhythms have and to '"keeps his [sic] ear open", but he does not only hear words, speeches, noises and sounds for he is able to listen to a house, a street, a city as he listens to a symphony or an opera. Of course he seeks to find out how this music is composed, who plays it and for whom' (Lefebvre 1996: 229).

The second process rhythmanalysis refers to is sensory rhythms. We can imagine this in terms of a visual rhythm, an olfactory rhythm, an aural rhythm, a tactile rhythm that mark the sensory landscape and experience of a particular place. These sensescapes fluctuate in intensity and in their relationships. Rodaway (1994) distinguishes five ways in which the different senses relate and are combined to generate specific geographical experiences or rhythms:

1. a *cooperative* relation, which means that senses operate together and thereby provide us with holistic information about the environment rather than partial information;
2. a *hierarchical* relation which refers to the domination of one sense over the others. Hence each modality of perception varies in its meaning according to the wider cultural belief systems that it is embedded in and the emphasis that a society gives to a particular sense. For example, a society might 'read' its environmental clues through their olfactory organs. Their landscape is therefore a landscape of smells in which the visual sense is relegated as being less important;

3 a *sequential* organization of senses, in which senses take a particular non-hierarchical order in our sensory perception and add an equal amount of information to our experience;
4 a relation in the form of *thresholds* where people experience degrees of environmental perception. Here senses come into play through different levels of stimulation. This can be culturally learned or is biological in origin and is also connected with the habituation to senses: for example, we perceive unfamiliar senses quicker than familiar ones;
5 a *reciprocal* relation, in which each sense establishes a link between the sentient and the environment and objects within that environment. Literature on the senses has largely focused on dialectical or non-reciprocal relationships between the senses. How the five senses are combined to produce particular place experiences has been largely neglected.

All of the above relationships are constantly occurring in different degrees of intensity in the city. However, at particular times different forms of relations predominate between the senses that influence the way we experience our environment. Thus 'in particular contexts, a certain sense and a specific style of operation of that sense (which is biologically and culturally determined) may play a hegemonic role in establishing geographical meaning' (Rodaway 1994: 37) and transform or redefine what constitutes our access to reality. We saw earlier how Simmel and Benjamin identified particular forms of interaction between the senses and forms of sensing with the emergence of modernity. It is a commonly held view amongst critics that every period in human history has its particular form of sensuous experience, in particular the emergence of urban-industrial societies is associated with major changes in sensuous experience and the redefinition of the senses (Schafer 1977; McLuhan 1962; Lowe 1982). This study seeks to determine whether similar changes are occurring in today's global postmodern society and illustrates how this affects the public life of cities.

Conclusion

This chapter has foregrounded the senses by anchoring them in daily geographical experience. This has been prompted by what I critique as an overemphasis on the visual in much literature on urban redevelopment. Instead I suggest developing a corporeal analysis by examining the *socially embedded aesthetics* to capture the multi-sensory experience of urban change. Let me briefly summarize the key parameters of such a sensitizing epistemology. This starts from the premise that experience is a corporeal process in which the senses work actively together. Senses order space and structure our 'sense of place'. Yet, the senses are not neutral but social in character. Different societies, different social groups and different historical times encourage particular forms of sensing and associate particular meanings and values with different senses, a theme I develop in more detail in the next chapter. Consequently, the same cityscape can be the arena where different sensibilities coexist and conflict, depending on the configurations that emerge between social and spatial forces, which structurally turn such spaces into spaces of encounter: that is, spaces of othering, crowding, fear, crime, surveillance and so on. The sensuous fabric of the city provides affordances that can ignite or diminish affective relationships between individuals and the sensed environment. A feature that I have described as 'place gestures' captures the particular atmosphere or mood that we associate with particular locations. To map this socio-cultural spatialization of places I suggested examining the sensuous and activity rhythms of an area.

Neighbourhoods that are experiencing intensified spatial restructuring in the form of urban regeneration schemes are particularly interesting case studies for analysing these links between the social and material world. The introduction of a new element in the urban environment is conducive to a range of transformations in the spatial experience of place in which certain 'rhythms' prevail over others, some disappear, new ones emerge and others continue simultaneously. Of course exactly what or who is seen, heard, touched, tasted and smelled is connected to questions about what is included or excluded in the experience of public space. This is an expression of power and the 'ability of certain groups to superimpose their rhythms on others' (Allen 1999: 65) as we will see in the next chapter.

4 Sensuous powers

In thousands of eyes, in thousands of objects, the city is reflected.
(Benjamin quoted in Gilloch 1997: 6)

Physical changes in the urban landscape produce new constellations of public spaces that reshape the sensuous, lived experience of place. Taking as a starting point Richard Sennett's remarks that '[d]iscussions of the senses lack much direct engagement in physical realities' (1998: 19), this chapter discusses the crucial role that senses play in framing power relations in the built environment. In our everyday encounters with the city, we tend to forget that the city is a human-made product and therefore reflects and expresses the values of a society (Lefebvre 1991; Markus 1993). Urban landscapes can be understood as places in which certain preferred meanings and practices are ingrained in their physical texture in order to maintain relations of power. In other words, cities are ideological constructions. Of course individuals can either go along with, ignore, or subvert these environments. Hence the organization of urban experience can be understood as a constant negotiation between an imposed order and individual agency. It is within this *space* of negotiation that power is felt, 'when someone acts in a way which they would otherwise not have done – regardless of whether or not they choose to' (Allen 2000: 10).

Power thus has always a spatial element. It is mediated through space and operates in specific places and contexts. Yet, simultaneously, power is a notoriously difficult concept because how it is expressed and how it operates can vary enormously (Lukes 1974, Allen 2003, Dovey 1999). Part of the difficulty lies in its ambiguous nature, which refers to both the capacity to control, as in 'power over', and the capacity to emancipate, as in 'power to'; hence 'one person's empowerment can be another's oppression' (Dovey 1999: 9). Let us take the example of graffiti in the city. A city authority might declare graffiti a criminal act. A graffiti artist decides to subvert this institutional power and make a drawing. Some passersby might ignore it, others will experience it as an act of vandalism and an imposition of subversive power. Yet others might read it as an act of empowerment and expression.

So how can we start to disentangle the multiple workings of power in the built environment?

My contention in this chapter is that power works through a network of associations between the material and social world in public space. Power is conceived here in a relational manner and as being constantly reconfigured. Modalities of power have become more dispersed and fluid, infiltrating the daily life of individuals in more complex and insidious ways. This has implications for how we conceive domination and resistance in space. In the last section I consider how the production of urban forms has historically been shaped by sensuous ideologies that aim to control order, purity and fear of the 'Other', which are still dominant in contemporary urban regeneration projects.

Flows of power

Public spaces are spaces of contestation in which different social groups negotiate their right to space (Mitchell 2003). Behind these contestations lie claims to power in space: whose practices, meanings and experiences are allowed and reflected in a place. As discussed in Chapter 2, the perceived spatial practices in a space are the outcome of negotiations between the conceived, the actual designed environment, and the lived – the 'imaginary geographies' of each individual. The conceived and lived dimensions tend to entail different sensuous perceptions of place, as Certeau elaborates in his analysis of the city from a bird's-eye view or from the street:

> To be lifted to the summit of the World Trade Center is to be carried away by the city's hold. One's body is no longer criss-crossed by the streets that bind and re-bind it following some law on its own; it is not possessed – either as user or used – by the sounds of all its many contrasts or by the frantic New York traffic. The person who ascends to that height leaves behind the mass that takes and incorporates into itself any sense of being either as an author or as a spectator . . . His attitude transforms him into a voyeur. It places him at a distance.
>
> (Certeau 1985: 123)

When looking from above, often the vantage point of the 'conceived', one gains an abstract view of the city in which the visual sense predominates and where one is distant from the viscosity of the urban. Yet, from the perspective of the pedestrian, one is enmeshed with the environment, 'possessed' by the sensory overload of the city. These two sensuous perspectives often do not coincide and produce conflicts over the practice, experience and representations of space. It is only by analysing the tensions and overlaps between these different sensuous perspectives of space that one can examine power practices in the built environment; as they are 'multi-dimensional, they cannot be simply addressed as forms of representation, lifeworld experiences or spatial structure; rather places are constructed, experienced and understood within the tension between these paradigms' (Dovey 1999: 3).

While urban semioticians (Gottdiener and Lagopoulos 1986) would argue that we can discover ideologies in the built environment by analysing discourses in brochures and interviews, Lefebvre (1991) suggests that much of this ideology functions at the practical level which people do not articulate verbally: 'The actions of social practice are expressible but not explicable through discourse; they are, precisely *acted* – and not *read*' (1991: 222, emphasis in original). Power in the city therefore needs to be studied as a mundane, almost banal characteristic which is expressed in and 'experienced' through the everyday: 'Daily life is the screen on which our society projects its light and its shadow, its hollows and its planes, its power and its weakness; political and social activities converge to consolidate, structure and *functionalize* it' (Lefebvre 1971: 64–5, emphasis in original).

But how do 'political and social activities converge' to consolidate, structure and functionalize daily life in public spaces? To understand the spatialization of power in public space, and its link to social relations, we need to analyse the relationship between flesh and stone, the human and the non-human.

Richard Sennett (1994), the most significant writer on the city and senses, examines the historical development of cities such as Athens, Rome, Venice and Paris and shows how the physical spatial order, social relations and the public imaginary of places are intricately linked by underlying sensory regimes. These sensory regimes divide, structure and order spaces in particular ways which influence our relationship with other people in places: 'The spatial relations of human bodies obviously make a great deal of difference in how people react to each other, how they see and hear one another, whether they touch or are distant' (Sennett, 1994: 17).

Sennett offers as an example for this urban ordering of social relations the Jewish quarter of Venice during the Middle Ages. Here the Jewish inhabitants were gradually segregated into a marginal area of the city – a ghetto – which involved other people 'no longer having to touch and see them' (Sennett 1994: 216). Jews were associated with polluting properties and, more importantly, with destabilizing the existing Christian order of the city. Sennett's (1994) argument is that urban spaces take their form from the ways that people experience their own bodies. I expand this line of reasoning by examining the role of the senses in producing or subverting spatial power structures in regenerated public places.

To understand how the senses play a crucial role in framing social relations, and hence how aesthetics are socially embedded, we need to take a closer look at what happens when our bodies are situated in specific places. Senses never work on their own, but need a reference provided by objects. We do not sense in a vacuum but need to be confronted with a material world to sense: for example, a flower we smell, a path we step on and touch, or food we taste. Objects afford certain sensescapes. Thus a constant interaction between our bodies and physical reality constitutes our experience of daily life. However, for most of the history of social theory, the role of things or objects in helping to constitute the social has been relegated to the background. Only recently have relations between humans and non-humans come the fore in feminist studies, cultural studies and actor-network theory (Latour 1993, 2000; Law 1986; Haraway 1991; Knorr-Cetina 1997; Urry 2000; Whatmore 2001). In the following I draw selectively on some of these texts to elaborate how power is mediated sensuously in public space.

Actor-network theory has developed a notion of society that highlights the 'human connectedness to the material world' (Knorr-Cetina 1997: 16) and incorporates the role of objects, in our case the built environment, in the creation of the social. We have to move away from the notion of the social as some hidden source of causality, and instead analyse how society and the social are 'composed, made up, constructed, established, maintained and assembled' (Latour 2000: 113). Accordingly, we need to examine how social and material processes become intertwined in complex sets of associations. This means that '[m]achines, objects and technologies are neither dominant of, nor subordinate to, human practice, but are jointly constituted with and alongside humans' (Urry 2000: 78). This involves a reworking of the notion of agency as a solely human characteristic, and to regard it 'as stemming from the mutual intersections of objects and peoples' (Urry 2000: 18). The experience of an environment is not merely a subjective practice but an active dialogical expression between the users of space and the possibilities that the constitution of place engenders in the form of sensuously perceived affordances. Such a view leads to a relational conceptualization of space in terms of 'networks'. This is a notion that captures the ways that spaces emerge as constantly reconfiguring socio-material relations rather than being arranged in fixed orders as in Cartesian space. The social world is constituted by hybrid relations, so that a public place is composed of a combination of human practices, physical material form and discourses. Hetherington describes this as 'the materiality of place' which 'does not mean that there is no space for the subject and subjective experiences and memories of space; rather they become folded in the material world and each becomes imbricated in the agency of the other' (Hetherington 1997: 184). It is precisely the senses which mediate the social and material bond in public spaces.

What are the implications of such a relational structure for our analysis of the sensuous framing of power? Rather than viewing power in this context as a tool in the hands of a few which is exercised *over* others (for a critique see Lukes 1974), or in a Weberian manner which regards power as a distinct form of coercion by individuals, I draw instead on Latour's (1986, 1993, 2000) relational conceptualization of power. He regards power as a process which, in order to be exercised, needs to be transmitted through a chain of people and objects who modify, change and add to it, according to their own purposes. Latour, drawing on Foucault, shows how the interrelation and associations between the human and material worlds provide the stable structure through which power is practised. He makes the 'material' side of power more explicit. This is a feature also emphasized in the work of Lefebvre (1976), who argues that power cannot be located in a specific site in space, nor is it in the hands of a few. It is rather part of daily spatial relations and interiorized in the individual through the interaction with, or sensing of, the material world:

> Power, the power to maintain the relations of dependence and exploitation, does not keep to a defined 'front' at the strategic level, like a frontier on the map or a line of trenches on the ground. Power is everywhere; it is omnipresent, assigned to Being. It is everywhere *in space*. It is in everyday discourse and commonplace notions, as well as in police batons and armoured cars. It is in *objets d'art* as well as in missiles. It is in the diffuse preponderance of the 'visual', as well as in institutions such as school or parliament. It is in things as well as in signs (the signs of objects and object-signs). Everywhere, and therefore nowhere . . . [P]ower has extended its domain right into the interior of each individual, to the roots of consciousness, to the 'topias' hidden in the folds of subjectivity.
>
> (Lefebvre 1976: 86–7, emphasis in original)

In Latour's view power is a fluid, mouldable process in which the initial 'message' is adapted and reinterpreted in different ways to serve people's various goals. Power here operates in a hybrid network of humans and non-humans and is therefore not uniform but experienced and expressed through multiple modalities. Let me clarify this. With regard to the urban environment this means that while the introduction of a new building alters the sensuous landscape, the dominant sensory experiences are not fixed, but are relative to the different users of place and the affordances the building allows for. For the purpose of my argument it is important to highlight that diverse individuals or social groups will experience and react differently to power. Some will be dominated, some will be oblivious, while others might directly subvert relations of power depending on their attachments, relationships and position vis-à-vis particular places. While certain sensory experiences and spatial practices might be designed into landscapes, these are always open to resistance or subversion, or might even be unnoticed. It is only through an analysis of specific settings and an examination of the interactions of the various players, from architects to residents or visitors, that we can uncover the micro-politics in urban regeneration processes.

The central issue that these various authors highlight is that power should not be viewed as a 'thing' that is possessed but as a fluid, relational process, expressed in a variety of forms that constantly need to be remade and negotiated in everyday life. This does not mean that power is also structured and centred at various points, such as through the state or through labour, and exercised in particular places. Sensuous power operates through the intertwining of the human and the environment or, in Sennett's words, of flesh and stone. It implies that power is not to be conceptualized as an abstract process in an analysis of 'socially embedded aesthetics' but rather in terms of a material expression. This certainly has to do with meanings and discursive processes but it also works through the hybrid relationship between practices and artefacts. This hybrid relationship is mediated through the senses. The relevance of this relational view of power is crucial in thinking through how power structures are being redefined in contemporary societies, which I discuss in the next section.

Control through pleasure

One of the key techniques through which power is interiorized is described famously by Foucault (1977, 1991) as discipline. Discipline is based on the constant surveillance of the body in space with the ultimate aim of order. These disciplinary techniques focus on individualized soul-training which was based on a systematic and generalized form of repression. Foucault focused his early research on the transformations of power relations since the eighteenth century, paying particular attention to how the Enlightenment, with the rise of modern institutions and scientific rationality, and the emergence of the nation state and capitalism, brought with it particular forms of domination. However, since then a number of changes have taken place in society which have altered the ways that power operates.

Globalization processes have challenged the experience of place and forms of communication that underpin conventional notions of place and time. For writers on postmodernity the disruptions of space and time in everyday life are beginning to take historical dimensions in which economic, political and cultural flows permeate and destabilize or reshape local–global relationships, creating different kinds of social relations (Lash and Urry 1994; Featherstone 1991; Harvey 1990; Castells 1996). For the current stage of the modern era 'flow' is increasingly becoming the leading metaphor (see Thrift 1996; Shields 1997; Urry 2000; Bauman 2000; Amin and Thrift 2002). To put it simply, this implies a move from solid structures that are bound in space and time to more fluid processes that trespass across time and space. Thus, while Foucault's legacy is important, the specific techniques of power that he described, such as the logic of the panopticon which saw power exercised systematically in a bounded place without an identifiable locus of power, is expanded by increasingly less tangible, more dispersed, mobile and fluid techniques of power. Foucault (1980, 1986) begins to address these issues in his later work, where he discusses the microphysics of power that do not require a central disciplining gaze. However, he pays little attention to the role of the senses in this new framework of power. In the post-panopticon stage, power is not restricted to a specific location or physical entity but by an absence of the locus of power: 'the prime technique of power is now escape, slippage, elision and avoidance' (Bauman 2000: 11).

This has led some critics to argue that we have moved from 'disciplinary societies' to 'societies of control' (Deleuze 1992, Rose 1999). Increasingly the state disperses control to other institutions whose practices filter down to everyday life. Hence, rather than individuals moving from one disciplinary institution to the next (from family, to school, to work), each seeking to mould our behaviour and encouraging practices of self-scrutiny and self-constraint, we are now in a society where control and discipline are designed into the flows of everyday existence (Rose 1999). There is a growth in regulating global flows and locally enforced mechanisms of control and discipline. Control and discipline work side by side, nevertheless expressed through different modalities of power. While the disciplinary regime attempted to alter individual behaviour and motivation, the new modalities of control alter the physical and social structures in which individuals behave (Rose 1999, Bauman 2000).

An example of this new form of power techniques in the built environment is Disney World (see Shearing and Stenning 1996). The source of its disciplinary regime, and minimization of disorder, is a mixture of physical and sensuous coercion, where 'every pool, fountain and flower garden serves both as an aesthetic object and to direct visitors away from, or towards, particular locations' (Shearing and Stenning 1996: 419). The environment induces cooperation and self-regulation in the consuming visitor. While the surveillance in this controlled environment is pervasive, it is distinguished from Foucault's notion of 'discipline' in that it does not require a detailed knowledge of the individual, and surveillance is installed both in the physical environment and in the social relations it aims to facilitate. Conformity is gained through pleasure. This 'control through pleasure' is based on a sensory manipulation of the environment. There are no unexpected sensory experiences that can shock, disgust or even alter the experience the person has in this environment; all the senses are 'in tune' with the hegemonic theme of the place. The experience of place is sensuously predetermined and prescribed. People's experiences are manipulated through a restrained sensescape. Commenting on the redevelopment of public places in New York, Zukin describes a similar 'pacification by cappuccino' (1995: 28) where the smell of cappuccino emanating from new bars is meant to linger, cover or replace other less desirable odours and attract customers. It reflects a hierarchy of senses in which one sense dominates the others. John Allen (2006) has defined this modality of power as 'ambient power'. In his study of commercial public spaces, such as the Sony Centre at Potsdammer Platz in Berlin, Allen identifies seductive spatial arrangements and ambient qualities as key to the workings of power. Power in these environments operates through experience:

> What goes on in such spaces, how they are used, is circumscribed by the design, layout, sound, lighting, solidity, and other affective means that have an impact which is difficult to isolate, yet nonetheless powerful in their incitements and limitations on behaviour.
>
> (Allen 2006: 9)

As opposed to Foucault's panopticon, there is no awareness of a guard watching individuals. Power here works through subtler means. People do not feel constraint by this environment, public spaces appear open and inclusive, yet their design and layout manipulates or silences and makes absent a variety of sensory perceptions (Allen 2006).

The absence or presence of certain sensory experiences has important consequences for power relations in public spaces. Lukes' (1974) reconceptualization of power relations provides an insight into how absence becomes ideological. He argues that decisions are made by individuals between alternatives. However, institutional practices can restrict the provision of choices and this might lead to individual inaction. Thus an institution might offer an individual a choice between options A, B or C. She might not want to choose any of them, but she is not offered the option of choosing anything else. Power here is interpreted in the ways through which individuals and groups 'through action or inaction, significantly [affect] the thoughts or actions of others' (Lukes 1974: 54). What interests me here is the emphasis on the role of inaction, silence, the unsaid, the not existing, or the role of 'absence' in the exercise of power, which prevents issues being discussed or the current state of affairs being taken for granted:

> [is it] not the supreme and most insidious exercise of power to prevent people, to whatever degree, from having grievances by shaping their perceptions, cognitions and preferences in such a way that they accept their role in the existing order of things, either because they can see or imagine no alternative to it, or because they see it as natural and unchangeable, or because they value it as divinely ordained and beneficial?
>
> (Lukes 1974: 24)

Most importantly this view of power highlights the significance of analysing the 'counterfactual', the absence of conflict, the non-event. Thus, an apparent case of acceptance or consensus in society can be interpreted as not being genuine but imposed. A poignant example is how regeneration discourses are portrayed as civic-minded because they physically and economically enhance an area and thereby try to suppress much criticism and opposition (Wilson and Grammenos 2005; see Balibrea 2001 for the creation of consensus in the 'Barcelona planning model'). Another example is the silencing of working-class, ethnic-minority and women's histories in most urban landscapes by not acknowledging or emphasizing buildings that reflect these groups' social history, or indeed by demolishing them (Hayden 1996, Purwar 2004): in other words, making them sensuously absent.

What I hope to have shown with this discussion is that less tangible, more dispersed and fluid techniques of power are emerging in the urban landscape. These new forms of power operate through strategies of control and alongside discipline by altering the social space in which individuals interact. This occurs by manipulating and silencing sensuous experience through a variety of modalities, from pleasure to absence. I now turn to explore how these power relations are negotiated in space.

Domination and resistance

So far I have provided quite a deterministic account of how power operates in present-day public space. However, public life is constantly reassembling itself as different groups and individuals access and appropriate it. Lefebvre's conceptualization of urban rhythms produced through sensescapes and activities captures some of these fluid formations. Thinking in terms of rhythms implies thinking about degrees of sensory intensity. As Lefebvre explains:

> It must be recognized that a deserted street at four o'clock in the afternoon has a meaning as powerful as the swarming square during those hours of trading or encounters. In music and in poetry, silences also have their meaning.
>
> (1996: 236)

By seeing social space as constituted through 'rhythms' Lefebvre draws attention to an important dimension of perception, namely that it is open to constant change, manipulation and alteration. In other words, domination and resistance in space always work side by side.

Lefebvre (1991) distinguishes between two styles of spatial power in contemporary societies: 'monumental spaces' and 'dominated spaces'. 'Monumental spaces' are spaces of a more traditional religious or political kind such as cathedrals or the Taj Mahal. While they certainly overpower the individual through the presence of certain sensescapes, they are based on people identifying with these places through a variety of sensory experiences that are set off by the spatial practices in and surrounding them. Monumental spaces work through gestures of place. Hence monumental space 'has a horizon of meaning: a specific or indefinite multiplicity of meanings, a shifting hierarchy in which now one, now another meaning comes momentarily to the fore, by means of – and for the sake of – a particular action' (Lefebvre 1991: 220). The sensory experiences in monumental places create a range of allegiances between the individual and the place. Yet 'monumental spaces' are increasingly disappearing from contemporary cityscapes. They are replaced by Lefebvre's 'dominated space': that is, one ruled by the hegemonic forces of capitalism. This is linked to Lefebvre's argument that we now live in 'abstract space' – space ruled by the logic of capitalism and implemented by a particular group of people: capitalists, bureaucrats and city planners (Stewart 1995). A particular order of space is imposed from above, this time without seeking consensus: 'In order to dominate space, technology introduces a new form into a pre-existing space – generally a rectilinear or rectangular form such as meshwork or chequerwork . . . Dominated space is usually closed, sterilized, emptied out' (Lefebvre 1991: 165). These 'dominated spaces', which are notably increasing in number around the globe, express their power through a particular sensory landscape that detaches the space from its surroundings. Power here does not operate through acceptance and identification with the existing locality but through alienation from it; their sensory reference group exists elsewhere. The present sensory regimes are absent from any local connection. An example would be the square surrounding François Mitterrand's Grande Arche at la Défense in Paris (Bauman 2000: 96). Its design and sensory landscape produce an inhospitable place that discourages staying: flat vastness, no benches, complete exposure to the weather, glass buildings that encourage one to look at them but not into them, and whose main entrances do not face outwards. The only movement is by the pedestrians who emerge from the metro exit and quickly disappear from view.

Opposed to the bleak view of 'dominated space' Lefebvre identifies an alternative space, the space of the communal, taken over through use over time, which he defines as 'appropriated space':

> a natural space modified in order to serve the needs and possibilities of a group that has been appropriated by that group. Property in the sense of possession is at best a necessary precondition, and most often merely an epiphenomenon, of 'appropriative' activity, the highest expression of which is the work of art.
>
> (Lefebvre 1991: 165)

An example of appropriated space would be the 'Reclaim the Streets' movement on 1 May 2000 in London when the statue of Winston Churchill (symbolizing traditional British values) was transformed for a few hours into a figure of anarchy, his head decorated with green grass in the style of a Mohican haircut, and blood painted coming out of his mouth. Domination of space is never an accomplished fact but relies very much on a temporal, and thus momentary, control of space. However, the body is a crucial element in subverting or bypassing dominated space as its sensory organs reframe or alter imposed meanings. The prescriptive space of a busy railway station, a space of transition, can suddenly become a rave club as young people plug in their iPods and dance to the latest rhythms (see *Evening Standard* 5 June 2007). This is Lefebvre's *total body* which resists through its sensuousness the imposed order and acts as a space of resistance, developing its own experiences:

> Thanks to its sensory organs, from the sense of smell and from sexuality to sight (without any special emphasis being placed on the visual sphere), the body tends to behave as a *differential field*. It behaves, in other words, as a *total* body, breaking out of the temporal and spatial shell developed in response to labour, to the division of labour, to the localizing of work and the spacialization of places.
>
> (Lefebvre 1991: 384, emphasis in original)

Certeau (1984) has similarly reformulated resistance in relation to the subversive powers of the body and the unpredictability of human behaviour. His work problematizes power structures in the urban environment and provides a model which emphasizes overlapping, heterogeneous layers of power in public spaces. For Certeau 'monumental spaces' or 'dominated spaces' exert their power through spatial strategies. Spatial strategies are those power relations imposed by the agencies in power, predominantly using permanent structures, such as metal barriers created to channel movements by commuters, or concrete elevations designed to slow or prevent access to buildings. These, however, can be constantly subverted through 'tactics', a range of non-planned activities, procedures or alternative experiences that do not have a sensed location but are activated through time and momentarily manipulate the terrain: 'strategies are able to produce, tabulate and impose these spaces, when these operations take place, whereas tactics can only use, manipulate, and divert these spaces' (Certeau 1984: 30).

Both strategies and tactics are deeply connected to the sensory experience of the place in the ways they establish what can be afforded in a place, and by presenting the environment as a resource for both domination and resistance: 'Disciplinary power *tells* us that a chair is for sitting on, but ecological perception permits us to see that it affords standing upon, throwing, lying over, scratching against, and so on' (Michael and Still 1992: 881, emphasis in original). However, while affordances can establish constraints and possibilities in an environment, it is down to the individual to activate these: '[a]ffordances provide a *resource*, not an inevitable *source* of resistance' (Michael and Still 1992: 883, emphasis in original). Senses can be part of a domination strategy, as illustrated by the square at La Défense, but they can also exert tactical powers, as, for example, in the way the shouts of children overpower the silence of a monumental space, or the smell of fried doughnuts from a van displace the cappuccino smell from surrounding cafés. The urban form is never static or complete but is constantly subverted by urban life (Delgado 1999).

Yet, when conceiving of resistance spatially, we should not view it only as a direct counter-space of 'oppression' but take into account the complex and varied forms through which resistance might be expressed (Pile 1997).

Certeau's vision of resistance through tactics highlights the fact that resistance 'seeks to occupy, deploy and create alternative spatialities from those defined through oppression and exploitation' (Pile 1997: 3). This is supported by my analysis of residence discourses and experiences of regeneration, where resistance is expressed in forms that bypass dominant meanings and instead produce different spatialities of experience. Hence residents often do not express a direct opposition to a regeneration process but subvert the imposed meanings by producing alternative spatialities that contradict or sidestep the regeneration discourses. These can range from a desire to experience non-regenerated environments, the consumption of marginality, or an intense dislike of certain groups of outsiders entering the area. In Pile's words:

> resistant political subjectivities are constituted through positions taken up not only in relation to authority – which may well leave people in awkward, ambivalent, down-right contradictory and dangerous places – but also through experiences which are not so quickly labelled 'power', such as desire, anger, capacity and ability, happiness and fear, dreaming and forgetting.
>
> (Pile 1997: 3)

An analysis of 'socially embedded aesthetics' has to take into account that this resistance operates not simply as a counterspace of oppression but creates its own spatialities of experience that are dis-located from spaces of domination (Pile 1997). Because of their transitional and fluid character, both in terms of spatial practices and sensory experiences, public places can never be completely claimed by power or subverted, only momentarily and partially.

Sensuous planning ideologies

> In the period of which we speak, there reigned in the cities a stench barely conceivable to us modern men and women. The streets stank of manure, the courtyards of urine, the stairwells stank of mouldering wood and rat droppings . . . the unaired parlours stank of stale dust, the bedrooms of greasy sheets . . . People stank of sweat and unwashed clothes; from their mouths came the stench of rotting teeth . . . and from their bodies came the stench of rancid cheese and sour milk and tumorous disease.
>
> (Süskind cited in Howes 1991: 145)

Süskind's description of eighteenth-century Paris illustrates two important issues. First, it makes a point against romanticizing intense multi-sensuous experience. Second, and more importantly, it allows us an insight into the changing sensescapes of public life over the ages, and our response of disgust illustrates how the reaction to sensuous landscapes is always linked to historically specific attitudes and social perceptions, a theme I discuss in this section. More precisely I consider the power of the senses in framing urban planning discourses and practices. I examine the sensuous shaping of three ideologies that I regard as pivotal in the transformation of public spaces in the past and the present: order, purity and control of fear.

Senses have a profound influence on social history (Corbin 1986; Classen *et al.* 1994; Synnott 1991). Let me clarify this claim by giving a brief account of the ways that smells have guided the development of the Western urban form. Despite the increasing dominance of sight in elaborating scientific discourse, Corbin (1986) shows that smell featured crucially in medical and public health theories of the eighteenth and nineteenth centuries. His account starts in the eighteenth century when it was believed that illness and death could travel through bad air, as putrid odour was considered to be the materialization of infection. Consequently ventilation, the creation of spatial distance and the uncrowding of places were regarded as ways of preventing illness. Naturally this approach to controlling currents of air also resulted in the deodorization of public space. In the nineteenth century a major shift occurred in the social imagination of smell which started to equate foul smells not only with death and illness but with filth, dirt, poverty and misery. Strong odours became associated with the lower social classes, adding a sense of disgust and a fear of illness. Quickly the working classes were associated with fetidity and dirt, 'the secretions of poverty' regarded as stemming from their social disorder and undisciplined behaviour. These perceptions of smell would deeply influence the spatial organization of cities.

During the nineteenth century a major concern of city planners in European cities such as Paris or Barcelona was ventilation. Old city walls were overthrown; hospitals, as well as businesses that caused unpleasant smells, were located outside the city; and roads were widened. City reformers supported the idea that sunshine and the circulation of air purified. So, for example, Haussmann's town planning could be regarded as aiming to eliminate the darkness of the centre of the city (Corbin 1986). Increasingly it became of interest to the ruling classes to spatially separate themselves from the unruly lower classes, who were perceived to live amongst, and emanate, strong odours. For the bourgeoisie the city was increasingly divided into places where they could give themselves over to the enjoyment of perception, where nothing offended the senses, as opposed to the 'non-city', the sensuously overwhelming and disordered spaces, where the dirty, the poor, the malodorous existed. Gradually smell defined the social order in the city and increasingly played a determining role in urban spatialization, as Corbin explains:

> Haussmann's policy could be interpreted – and not unreasonably – as a 'social dichotomy of purification' . . . social division of stench – almost uniformly distributed in the not so distant past – was now in force in the

> city . . . it was thought that purification requirements must be selective, quite apart from the fact that disinfecting the space reserved for bourgeois activities could only enhance property values. Wealth increased when the volume of refuse and the strength of its stench decreased. On the other hand, purifying rented premises crowded with apathetic workers would do nothing for the time being but add inordinately to landlords' expenses. The quest for profit strengthened this social distribution of odours.
>
> (Corbin 1986: 134–5)

As this quote highlights, the smellscape of places and their associated social imaginary determined the economic value of places. In other words, the salubriousness of places was determined by economic interests, which is still a dominant feature in contemporary urban regeneration projects.

This short account highlights three key points for the development of a 'socially embedded aesthetic' analysis of public space. First, senses are never neutral. Their meaning and what emphasis we lay on each individual sense is shaped by society. We can draw from this that the 'reality' a culture accepts is closely related to the ways it defines its sensuous experience and can therefore conclude that senses play a hegemonic role in establishing geographical meaning (Rodaway 1994). Second, giving priority to a certain sense, as shown in the above example, is linked to a particular worldview and experience of space. As argued in the last chapter, each sense establishes a particular relationship with the environment. Following Corbin's (1986) argument, we can appreciate how a focus on smells informed a public discourse concerned with draughts, circulation and air which was reflected in the spatial practices of urban planners. Third, the behaviour of the upper classes in the nineteenth century shows how our expectations of public space and fostered spatial practices are closely bound to the sensuous geography established by each society. Sensuous geographies are not 'natural', but discursive constructs which inform our views and relationships with particular places and are negotiated in everyday life. Context-bound behaviour is linked to certain presuppositions and norms about what kind of sensuous experiences are allowed and expected in public. Each society constructs certain assumptions about what it is permissible to see, taste, hear, feel and smell in a public space. We can conclude that sensuous geographies are important elements in the construction and maintenance of *social order* in place.

This leads me to the second ideology informing urban planning: purity. What determines the right 'order' in cities? How do we differentiate between what is considered ordered and disordered? Inherent in the notion of order is, as the anthropologist Mary Douglas (1966) poignantly argues, the notion of 'purity' or cleanliness. Hence what makes spaces ordered is that they are regarded as pure:

> dirt is essentially disorder. There is no such thing as absolute dirt: it exists in the eye of the beholder. If we shun dirt, it is not because of craven fear, still less dread or holy terror. Nor do our ideas about disease account for the range of our behaviour in cleaning and avoiding dirt. Dirt offends against order. Eliminating is not a negative movement, but a positive effort to organize the environment.
>
> (Douglas 1966: 2)

Douglas' definition of dirt links sensuous experience, mainly visual and tactile, with the environment. By eliminating or moving around unwanted sensuous experiences we order our surroundings, we construct 'pure', clean environments. The notion of place itself always already presupposes an established mode of classification that defines the 'right order', 'a sense of the proper', of '[s]omething that *belongs* in one place and not another. What one's place is, is clearly related to one's relations to others' (Cresswell 1996: 3, emphasis in original). Thus, implicit in schemes of perception which 'order place' are intrinsic sensuous ideological dimensions, as they delineate what is the right order, what is pure, what is impure – what is considered 'dirt'.

Sibley's (1995) study of 'geographies of exclusion' illustrates how associations, images of places and the perception of their inhabitants are informed and constructed by broader cultural discourses of purity which lead to subtle spatial exclusionary practices, and are shaped by those in power. By drawing on object relation theories, he shows that to develop a balanced personality every individual has the tendency to reject difference and search for order. This tendency is reinforced by institutional controls and broader social discourses that exclude 'matter out of place' (Douglas 1966). It is the latter point which this study aims to analyse in regard to the redevelopment of contemporary urban landscapes. For Sibley (1995), exclusionary practices are often based on particular notions of purity that regard dirt as imperfect and inferior:

> Exclusionary discourse draws particularly on colour, disease, animals, sexuality and nature, but they all come back to the idea of dirt as a signifier of imperfection and inferiority, the reference point being the white, often male, physically and mentally able person.
>
> (Sibley 1995: 14)

Referring back to Douglas, Sibley argues that the people, animals and things that do not fit into a group's classification scheme are those that are considered 'impure'. In other words, they are polluting the environment through their mere presence, secreting sensory pollution. People themselves are regarded as 'carriers' of threatening disorder or acting as mobile 'dark spaces'.[1] This explains why much city planning involves the control of spatial movement and spatial segregation. Social hierarchies are played out in space through the employment of negative or positive sensory associations which define who and what is pure or impure. This suggests that sensory associations or descriptions are techniques of power which determine who is allowed or is barred from a place. Hence attempts at ordering the city can also be interpreted as forms of purifying the city, designed to exclude groups identified as sensuously polluting.

Moreover, as Stallybrass and White argue, oppositions of order–disorder, purity–impurity are not only contrived but necessary for the dominant culture to establish themselves as superior and different from the 'Other': 'what is *socially* peripheral is frequently *symbolically* central' (1986: 5, emphasis in original). Matter out of place is far from residual, but a necessary part of the shared imaginary repertoires of the dominant culture. Normative discourses can only exist when they can differentiate themselves from the deviant: 'the greater the search for conformity, the greater the search for deviance; for without deviance, there is no self-consciousness of conformity and *vice-versa*' (Davis quoted in Sibley 1995: 39, emphasis in original).

The city is a place of 'exposure' to the Other (Sennett 1990), to the unknown and unfamiliar. However, relations in the city are permeated and shaped by social hierarchies. Anonymity, the exposure to the stranger, the unknown, rather than being regarded as positive elements in the city are most of the time linked with feelings of fear. To speculate about the origins of this fear would be outside the remit of this book, but I would like to sketch out the third ideology that has shaped the sensuous landscape of the city: the control of fear.

The work of Ellin (1997a, 1997b) analyses how the form and planning of the city are shaped by notions of fear and safety in society. He illustrates how from antiquity to the Renaissance the city was conceived as a 'safe place', as exterior walls protected the citizens from outside invaders. Since the advent of industrialization, however, and changing social, economic and political circumstances, the city itself has been increasingly associated with internal dangers. This is linked, as Ellin (1997b) suggests, with broader changes and feelings of insecurity that industrialization and the move to modernity brought, where traditional structures and institutions start to stumble and, as Berman (1983) citing Marx states, 'all that is solid melts into air'. Hence, not surprisingly, the beginning of the nineteenth century witnessed the start of zoning – an attempt to rationalize and order space socially by determining the nature of the built and spatial form. Rationalization had inevitable effects on the character of public space, which became more functional as streets were changed from being spaces for long encounters to become places of flow and movement; '[t]he social and productive aspects of the street and market place particularly were suppressed in favour of movement and consumption respectively' (Ellin 1997b: 21). Functionalism would achieve its heyday in modernist city planning, when increasingly the market shaped cultural forms of expression. Thus 'much of what was built after the war in the US . . . consisted of isolated towers and slabs as well as unending blocks of mass produced individual houses' (Ellin 1997b: 25). This emphasis on physical isolation or 'decorporealised space' (Diken 1998) can be interpreted as an ambition of modernist planners to neutralize strangerhood and exposure. The city is viewed as a space that should discourage close physical encounters with strangers: 'The decorporealised space described by a fear of touching is a space dominated by the eye, where the body and tactile reality are extinguished by the dominance of visuality' (Diken 1998: 72).

Recent writings (Ellin 1997a, 1997b; Davis 1999) seem to suggest that as much as society is transforming, fears are changing too. This 'postmodern fear' (Ellin 1997b) needs to be understood as a response to the acceleration of social changes since the 1960s, and the rise of what have been described as postmodern insecurities.[2] Linked to the emergence of a 'control society', postmodern fear suggests that danger can no longer be contained in a specific place. It cannot be located and 'disciplined', but is everywhere. The individual has to protect herself against constant potential danger. How does this 'postmodern fear' relate to the sensuous shaping of regenerated urban environments?

Two aspects are particularly relevant: the militarization of landscape and the use of history to regenerate neighbourhoods – polar opposites that reflect the workings of discipline and control through pleasure. The assurance of safety has become one of the most important features of contemporary regenerated spaces (Atkinson and Laurier 1998). As Davis (1998, 1999) shows in the example of Los Angeles, design, architecture and the police apparatus merge. Designers and architects are becoming increasingly experts in security, or maybe they always have been, as my discussion of Paris suggests. As Oscar Newman's *Defensible Space* (1972) illustrates (and this is still one of the best-selling book amongst planners[3]), it is generally believed that fear can be designed out. In his view safety can be provided in urban environments by creating real and symbolic barriers and by zoning the space into clearly defined private, semi-private and public spaces. In this way opportunities are provided for surveillance, either in the form of windows or, more recently, CCTV cameras. Safety is achieved by manipulating the design of objects in a particular space to define and protect the boundaries of an environment, so that it does not afford unwanted practices and experiences to socially marginal groups. We do not have to go to Davis's (1998) account of Los Angeles to observe how Newman's views are applied all over the globe, as old-fashioned benches that afford extended seating and space for lying down are replaced by barrel-shaped benches such as those at London's bus stops, or individual seats such as those in most of Barcelona's new public spaces. I will consider in the latter part of this book whether the same holds true in the regeneration of Castlefield and El Raval.

A less obvious way of combating postmodern fear, and strongly related to the policy of control through pleasure, is the use of heritage and the recycling of past urban forms in regenerated areas. As Ellin (1997b) explains, this is linked to escapism and nostalgic visions of a 'golden age' in which fear is not perceived to exist:

> Whereas modern fear and the positivistic climate in which it occurred led to efforts to detect cause and effects – to rationally understand the present in an effort to guide the future – postmodern fear amid the reigning anti-technocratic climate has incited a series of closely related and overlapping responses including retribalization, nostalgia, escapism, and spiritual (re)turn.
>
> (Ellin, 1997b: 26)

The reasons why historical environments or heritage are used to give a sensation of security are various. The aim of urban heritage schemes is a particular reframing of urban reality (Boyer 1992). Historic monuments are renovated, sandblasted and stripped of the sensuous indicators of age such as dirt, moss and traces of environmental pollution in order to fit in with new buildings and the contemporary character of the area. They are all integrated into a 'designer heritage aesthetics', as I will discuss in later chapters. The immediate present is ignored, as are the uncomfortable realities of the past. In this way both past and present problems are simultaneously erased and ignored in the idealized city tableau. The past is incorporated into a non-threatening and sanitized vision of the present. These regenerated environments plan fear out, as much in terms of their use, architecture and sensuous experience as in terms of the people they attract, reflecting Allen's (2006) argument about ambient power. People, so developers argue, feel safer with like-minded people (Davis 1998). Most of the time this like-mindedness is defined by people's consumption practices and expendable income. Non-consumption is constructed as a form of deviance. However, it is often not clear what the fear is about – an imagined Other? It is simplified into a generalized social perception of threat that becomes the function of security mobilization[4] (Davis 1998) and 'crowd control', which is achieved by more obvious techniques such as a direct denial of access to public space through private security guards or by homogenizing the space through the particular organization of sensescapes to make certain groups feel unwelcome:

> Ultimately the aims of contemporary architecture and the police converge most strikingly around the problem of crowd control. As we have seen, the designers of malls and pseudo-public space attract the crowd by homogenizing it. They set up architectural and semiotic barriers to filter out 'undesirables'. They enclose the mass that remains, directing its circulation with behaviorist ferocity. It is lured by visual stimuli of all kinds, dulled by musak, sometimes even scented by invisible aromatizers. This Skinnerian orchestration, if well conducted,

> produces a veritable commercial symphony of swarming, consuming monads moving from one cashpoint to another.
>
> (Davis 1998: 257)

What this quote from Davis highlights is that sensory experiences of places are deeply intertwined with values associated with particular social groups using this space, and this leads to a reinforcement of place identities. While he criticizes this orchestration of public space from an economic point of view, Sennett (1986) regards the consequences of homogenized spaces as even more detrimental to civic culture. Public space, rather than being a space of encounter and negotiation of difference, becomes a space where people and groups withdraw from society and create territorial barricades (both in white suburbia and in the inner-city 'ghetto'). Ironically, the outcomes are similar at both ends of the spectrum:

> The defended neighbourhood is characterized by a homogenous social group exerting dominance within its boundaries in reaction to perceived threats of territorial violation by outsiders. Street gangs use spray paint while homeowners' associations use neighbourhood watch signs, either way we are talking informal militias.
>
> (Flusty 1997: 57)

Conclusion

Power relations in urban environments are expressed, mediated and experienced through the senses. The construction of urban place is not only a geographical, political or economic matter but intersects with socio-cultural expectations, in particular with social perceptions of the senses. Conceived space, spatial practices and lived experiences are ingrained by the senses and inform each other. Sensuous ideologies materialize in the built environment. Drawing together the key arguments, let me summarize what conclusions we can reach from my discussion so far.

First, senses mediate the relationship between the material and the social and have structures of power ingrained in them. The senses do not neutrally mediate the material world: that is, the senses are not a transparent medium, but are shaped by specific ideologies in society. These ideologies influence not only how people perceive themselves and others in space, but also what an environment signifies to them. Yet the experience of space is based on ephemeral structures that constantly change. For an analysis of sensuous power this means that power relations in public space are fluid-relational in character and are configured through the association of various elements: discourses, practices, imaginations and experiences. A socially embedded aesthetic analysis aims to disentangle how this material side of power is constructed and exercised.

Second, I have established that power relations in today's 'control society' are negotiated through networks of power. Power no longer works by altering individual behaviour but instead by altering our everyday surrounding physical structures. Control and discipline are now designed into flows of everyday existence. At the same time, places are always open to resistance as the sensuous body produces different experiences and meanings, not necessarily by opposing the dominant sensescapes but by creating its own alternative spatialities.

Finally, city planning has historically been informed by sensory regimes that structure, order and divide the city to serve particular interests. Senses play a determining role in establishing these socio-geographical meanings and social hierarchies in the social production of urban space. Taking these points together has consequences for thinking about city life. A sensory analysis of public space and life is paramount for understanding the emergence of a new spatial order in the city.

Part II

5 Castlefield and El Raval

In 1999 for the first time in its history the Royal Institute of British Architects awarded the Royal Gold Medal for Architecture to a city, Barcelona. Almost in parallel, Manchester's urban renewal has been given increased attention around the world as Manchester has 'progressed to become Britain's most successful regional city, drawing international investors, creating thousands of new jobs and pioneering city-centre living' (Hetherington 2006). After years of oblivion, Manchester's and Barcelona's post-industrial cityscapes are celebrated as shining examples of successful urban regeneration and stylish public space design (Garcia-Ramon and Albet 2000; Quilley 2000; Marshall 2000). Similarities can also be found in the cities' history. Both were the cradle of the industrial revolution in their respective countries, Manchester being famously known as 'Cottonopolis' and Barcelona as the 'Manchester of the South'. The two metropolises developed a strong working-class identity reflected in their history of socialist local councils. Comparable patterns of deindustrialization affected their city centres in the 1970s and both cities resorted to similar culture-led regeneration schemes during the early 1990s to reinvent their physical landscape.

Against this background, I analyse how regeneration has reconfigured public space and life in the two neighbourhoods that have been most dramatically altered in the cities' new urban landscape: El Raval in Barcelona and Castlefield in Manchester. For the remaining chapters, I explore the role of the senses in shaping planning processes and informing spatial practices embodied in both the production and daily uses of public space. But first, to 'explore the continual dialectic between past and future forms' (Abu Lughod 1999: 3), we need to situate the two neighbourhoods within the broader spatial and social context of their respective cities. This chapter therefore starts with a comparative sensuous-historical evaluation of Castlefield's and El Raval's spatial, material and symbolic development. This is followed by an overview and analysis of their regeneration policies since the 1980s, focusing particularly on how the remodelling of their public spaces has been justified and put into effect by drawing on sensory paradigms.

A sensuous history

El Raval and Castlefield were historically characterized by two opposing and apparently contradictory features: first, centrality in terms of their geographical position in the city centre and their proximity to trade and transport networks. In the case of El Raval this was the sea, in Castlefield's case, the rivers Irwell and Medlock. Their second common feature – marginality – resulted from their dense urban development and their role as recipients of the city's residual activities.

Arguably the birthplace of Manchester, Castlefield took its name in Georgian times when the ruins of an old Roman fort (AD 43) overlooked the agricultural area outside the city. The transformation of Manchester from a medieval township into a thriving industrial city emerged simultaneously with the development of Castlefield into the city's central trading space. At the start of the nineteenth century a range of canal systems and large warehouses were built in Castlefield's basin area connecting it with the main canal transport routes of the north of Britain. Manual labour from all over the north-west was attracted to Castlefield and soon its population rose from virtually zero in 1750 to some 50,000 people in the nineteenth century.

Castlefield was characterized by a varied urban landscape containing a mix of factories, warehouses, commercial venues and rapidly built working-class housing, lacking basic sanitary facilities. The south-western rim developed into a meat distribution area housing abattoirs and skin yards where the smell of putrefying meat attracted rats, and disease lingered day and night: 'It stank to high heaven. The carcases were salted and left on the floor and there were bluebottles by the thousands!' (resident, quoted in Heaton 1995: 58). Working-class families as well as independent craftsmen such as bookbinders, tailors and clothmakers mixed on the bustling Liverpool Road, the main route crossing the neighbourhood.

Two iron markets were constructed in 1880, and are still standing today. Markets and fairs in Castlefield were major events in the town's recreational calendar until after World War II, attracting a variety of social groups. Castlefield 'was visited by thousands who otherwise would have had no reason to visit this dreary quarter in Manchester' (Brumhead and Wyke 1989: 27). Its negative reputation was an outcome of its working class and carnivalesque character that had fostered the emergence of marginal businesses: prostitution, located mainly on Tonman Street, and petty theft (Jones 1997). We have to imagine Castlefield during this time as a densely built and crowded place where intense visual, aural and olfactory experiences and close tactile encounters of a jostling crowd would quickly envelop the pedestrian. Yet, for much of the bourgeois population of Manchester, Castlefield remained an 'invisible place', physically and sensuously segregated from the city centre:

> [t]he inner and outer rings were joined by main thoroughfares along which a 'good' class of shops developed so as to service the bourgeoisie on their way to and from work. In the process, the working class housing 'disappeared' behind a respectable front.
>
> (Engels quoted in Stallybrass and White 1986: 130)

This zoning of cities was linked to the bourgeois sensuous regulation of the city. It was based upon the bourgeois imaginary of the working classes as a 'sanitary and moral danger',[1] which led to the segregation of whole areas of the city and stigmatized these places and their residents (Stallybrass and White 1986). Adding to its already marginal character, a number of hospitals opened in Castlefield during the nineteenth century such as the Royal Eye Hospital, the Skin Hospital, and the Children's Hospital, amongst others. Most locals did not approach the hospitals for fear of infection, a perception projected onto the whole neighbourhood: 'We knew about the place, but we didn't know too much about what it was for, and we wouldn't walk near or touch the building in case we caught it!' (resident, quoted in Heaton 1995: 3). We can see here how fears in the city, the shunning of certain areas and the creation of a spatial order, are all linked to fears about sensory pollution, especially transmitted through touch and smell.

With the opening of the world's first railway passenger station on Liverpool Road in 1831 the social, spatial and visual structure of Castlefield was radically altered. Various high-level viaducts were built from Central Station (1879) and the Great Northern Goods Depot (1894) right through the middle of the old Roman fort. This led to the rehousing of many working-class families to other areas of the city. Most notably, in 1877 the Southern Iron Viaduct was erected by the Cheshire Lines Company, and is still a prominent feature in the basin today. Castlefield was at this time filled by the roaring noise of trains passing, its streets bustling with passengers getting to and from the station. With the gradual decrease of industry and the partial destruction of the area by the bombing during World War II, Castlefield increasingly lost its importance as a transport centre. In the late 1960s Central Station closed and Liverpool Goods Station followed in 1975. The last working-class houses – by then referred to as slums – were cleared to make room for Granada Studios. Many of Castlefield's canals were filled with rubble and the abattoir was moved to another part of the city.

From being a crowded and dynamic working-class industrial neighbourhood, Castlefield deteriorated in the second half of the twentieth century into an abandoned place, devoid of most of its population; filled with scrapyards and derelict canals, its disintegrating industrial landscape was a symbol of decline as in many northern British cities (O'Connor and Wynne 1996). This decay was reflected in its sensuous geography:

> Castlefield is a decaying little known backwater on the fringe of Manchester's city centre . . . Like a doormat trodden on by the passage

> of time, it lies now in Manchester's forgotten no man's land – a city's classic backyard. There are places where the sun never shines beneath the stairway to the city's history. It is pitted by crofts, crumbling buildings and silted waterways and the weeds sprout up everywhere.
>
> (*Manchester Evening News*, 29 October 1979).

During the Middle Ages El Raval (meaning the 'periphery') was an agricultural zone squeezed in between two city walls.[2] Its spatial segregation from the city was accentuated by a sensuous frontier, a dirty sewage stream difficult to cross which is today's world-famous Ramblas.[3] Its geographical distance from the main city was paralleled by social marginalization as religious institutions and hospitals looking after the ill and poor settled in its northern area. With the advent of the industrial revolution in the nineteenth century, the southern side of the neighbourhood became built up with large factories and cheap, insalubrious working-class housing. As El Raval's industrial growth[4] progressed within medieval walls, living conditions were some of the worst in Europe and led to an oppressive yet vibrant atmosphere on its streets (Hall 1997). Nevertheless, some industrialists chose to settle in the bourgeois residences on the northern side of El Raval. The social divide between southern and northern Raval is still perceptible today in the physical structure of the place. Historical buildings and most cultural venues are located in the north. Southern Raval is considered its poor neighbour. The dividing line is an old Roman road, Carrer Hospital.

Only in 1859 were Barcelona's city walls demolished as its urban expansion started with Cerdá's grid-like 'Eixample'[5] (literally: extension). The spacious new apartments attracted the middle and upper classes who left the Old City. The power of bourgeois spatial ordering becomes once more evident as El Raval disappeared behind a new wall of so-called 'panel buildings'. These were built along the circumscribing avenues to hide the existence of a poor and degraded neighbourhood, and increased its physical and psychological isolation from the city centre. At the end of the nineteenth century, the industry gradually moved out in search of more space. The left-over factory shells would soon be filled by the first wave of immigration from rural Catalunya. Migrants settled here to work on Barcelona's first World Exposition in 1888. The continuous waves of immigration into this already populous area led to a flourishing of non-licensed residences and subletting, and to the construction of vertical shacks on most rooftops, so called 'barraquismo vertical'.[6] Small artisan industries such as printing, carpentry and distilling replaced the large-scale industry and have stayed in the neighbourhood until today. El Raval's urban texture was defined by a multilayered, complex and diverse structure which nowadays still gives the impression of a chaotic unplanned space.

One cannot talk of El Raval[7] without mentioning the name which would infamously describe the area during most of the twentieth century, namely 'Barrio Chino'.[8] The neighbourhood's working-class and industrial character was accompanied by prostitution and related sex industry businesses, mainly situated in its southern part, close to the harbour. Hence this zone was full of cafés, taverns, theatres, music halls, brothels, restaurants and other entertainment businesses catering for the upper classes and the poor alike. El Raval's 24-hour bohemian[9] and cosmopolitan feel became legendary and was soon compared to Paris's Montmartre. Streets were lined with flashing neon signs, punters would interact with prostitutes, and brightly lit shop windows displayed condoms and all sorts of sex toys. Villar (1996: 105) evokes its atmosphere:

> The street Nou de la Rambla was unique in the world. The gallant life, intense as ever, was concentrated on that street. The animation did not stop for a single minute. Afternoon and evening were connected, the evening with the night and the night with the morning. It possessed this universal character, as much appealing as disturbing. No person was considered a stranger; nobody turned their heads after the most shocking individuals . . . This particular idiosyncrasy created a legend.

The other side of this romanticized version of El Raval was its lived reality. In 1932 the density of the place was more than 1000 people per hectare and hostels and lodging houses were so overcrowded[10] that a strong odour emanated from their doors. Meanwhile the northern part of El Raval was a quiet area, at the margins of the colourful south. Hospitals, workhouses and orphanages provided shelter for the unwanted, and cloisters provided a Catholic education for young bourgeois women.

El Raval's bohemian and permissive character operated in opposition to the bourgeois Barcelona, and was politically further accentuated by the neighbourhood anarchist and trade union activity at the start of the twentieth century. A number of important protest marches against factory owners started here, and in 1936 armed trade unionists swarmed out from El Raval to try and crush fascism. The reason for El Raval being a container for 'marginal' activities such as prostitution, petty delinquency and anarchism – and therefore a sore in the eyes of the authorities – is connected to the spatial practices afforded by its multilayered urban structure. People could hide, run away, or shelter in the labyrinth of narrow streets and courtyards; many hid in corners and alleys or quickly crossed the neighbourhood along roof tops. In both the physical and symbolic geography of the city, El Raval was a disordered place where any means of control was difficult to establish. There were innumerable times when it became a no-go area for the police after dark (Villar 1996). We can see here how the public imaginary of a place is supported by its sensuous geography.

After the Spanish Civil War (1936–9) a number of factors led to the gradual fall of the Barrio Chino. In the aftermath of the war Catalunya, and especially Barcelona, suffered under Franco's repression by a conscious lack of government investment in public services. The imposition of a new moral climate led to the closing of theatres, brothels and other leisure establishments. These activities then went underground leading to a steady increase in street soliciting.[11] In the 1960s new social trends moved the nightlife gradually to other parts of the city. El Raval gradually became an area for cheap prostitution and sordid sex shops, segregated from the rest of Barcelona.

The exodus of residents from El Raval started in the late 1960s.[12] El Raval's constant role as a space of transition for newcomers to the city meant that during the 'Spanish boom' 60,000 residents left (Gabancho 1991). The final nail in the coffin was the advancement of democracy and the opening of Spanish frontiers. Hard drugs such as heroin entered El Raval and led to brutal delinquency, causing a further flight from the neighbourhood. In these years (1975–92) urban insecurity of both residents and outsiders increased dramatically, as the 'Barrio Chino' lost its frontiers, crossed Carrer Hospital to the north of El Raval and to the Ramblas and affected Barcelona's tourist trade. Those who could not leave were not only the poor, or those that had fallen through the social safety nets, but also respectable residents that had lived there for generations and refused to leave 'their' neighbourhood. El Raval's strong social cohesion, still perceptible today, stems precisely from its poor economic circumstances that led to low mobility patterns, dense living conditions and the emotional ties caused by generations of families living there, which have created a strong popular culture (Tatjer i Mir and Costa i Riera 1989; Garcia 1998).

If we compare Castlefield's and El Raval's historical development from a sensuous-spatial perspective we can reach the following conclusions. First, their densely built and inhabited industrial working-class urban landscape sets them physically apart from the rest of the city. Second, during the nineteenth century their activities situated them as marginal and carnivalesque spaces in the public imagination: on the one hand, as *places of disease* where non-desirable activities and institutions settled such as hospitals, skin factories, infant houses; on the other hand, as *places of diversion and eroticism* renowned for their entertainment venues and sexual decadence. These imaginaries are reflected in early twentieth-century paintings and prints[13] that depict both areas as bleak industrial places filled with streets of foul-smelling smoke from factory chimneys and where the noise of machines never abated. At the same time photographs, historic documents and novels depict a gregarious and busy street life produced by high levels of working-class population and poor housing conditions, which did not invite residents to stay inside. This fostered the growth of small businesses, shops, sex-related businesses and markets which encouraged a rich 'ethics of engagement' between different social groups.

In the second half of the twentieth century El Raval and Castlefield evolved into neighbourhoods whose sensescapes reflected decay. They become *places of loss*, however marked by different attributes. Before the regeneration process started in the late 1980s, while still densely inhabited,

El Raval became a socially marginal neighbourhood, its public spaces menacing because of drug-dealing and related criminal activities reflected in 'closed metallic shutters of bars and commerce, and the abundant flyers advertising flat rental, and in the hostels that were working as "hot beds", so-called because clients took turns in the rooms, 24 hours a day' (Villar 1996: 230). In the eyes of the city authorities El Raval was a human wasteland. Castlefield, on the other hand, had deteriorated into an industrial wasteland. An abandoned place, with hardly any residents, no activity on its streets, crumbling paths and buildings, the only sound was that of guard dogs, its air filled with the stale smell of canals and rotting wood.

By the 1980s Castlefield's and El Raval's early historical and geographical importance had seemingly been forgotten and their former centrality was no longer reflected in their comparatively poor integration into the city. Hence, just before their regeneration in the late 1980s, the spatial organization of the neighbourhoods reflected a poor 'economy of access' for outsiders: Castlefield because there were no public pathways, El Raval because outsiders did not dare to enter the labyrinthine streets filled with potential dangers. Those that managed to enter were faced with oppressive sensuous gestures created by squalid conditions, poor lighting and putrid odours. As I discuss in Chapter 6, one of the main aims of the regeneration was to reverse the physical and symbolic confinement of these spaces and open up the neighbourhood for the rest of the city and visitors. For the rest of this chapter I situate Castlefield's and El Raval's socio-spatial transformation within the urban regeneration trajectory of their cities. The focus will be on the years between 1975 and 2002 and will pay particular attention to the impact urban policies have had in the remodelling of public space.

Castlefield: from industrial revolution to inner city cool

The deterioration of Castlefield coincided with a general decline in British urban life after the 1950s and a trend of moving to the suburban outskirts of the city. Manchester city centre alone lost 180,000 residents between 1961 and 1976 and the city became one of the poorest areas in Britain (*Manchester Evening News*, 15 September 1978). Castlefield's rediscovery in the late 1970s has to be attributed to the relentless campaigning of local archaeological groups and historical societies[14] which fought hard to get the area's historical importance recognized (Schmidt 1994). Its 'existence' was only acknowledged in the public imaginary when the Department of the Environment designated the district an Outstanding Conservation Area in 1979 (this status has since been abolished). It is the combination of efforts by local conservation groups eager to preserve heritage and the orchestrating role of the local state that, as Urry (2002) contends, sparks the refurbishment of an area and provides a sense of local pride, as this newspaper article suggests:

> Today it is a grimy back alley of the city's past, tomorrow it could be Manchester's South Kensington. It's not the product of a planner's rose-tinted dreams either, but a possibility well on the way toward fulfilment as part of Manchester's historic awakening . . . In a report on Castlefield local historians say: 'The area should be developed as an open museum of Manchester's historic and industrial past to demonstrate the important role this city has played in the country and the North West in particular'.
>
> (*Manchester Evening News*, 31 October 1979)

Castlefield's 'historic awakening' involved a cultural resignification in which history was essentialized into specific spatial and architectural patterns to provide the area with a new role in the economic life of the city centre (Figures 5.1 and 5.2). The recoding of Castlefield was formalized in 1982 in a Tourism Development Plan. It proposed the transformation of Castlefield into Britain's first self-nominated Urban Heritage Park, a concept originating from the United States, and more precisely, in Lowell's National Historic Park in Connecticut (Schmidt 1994). This designation provided a

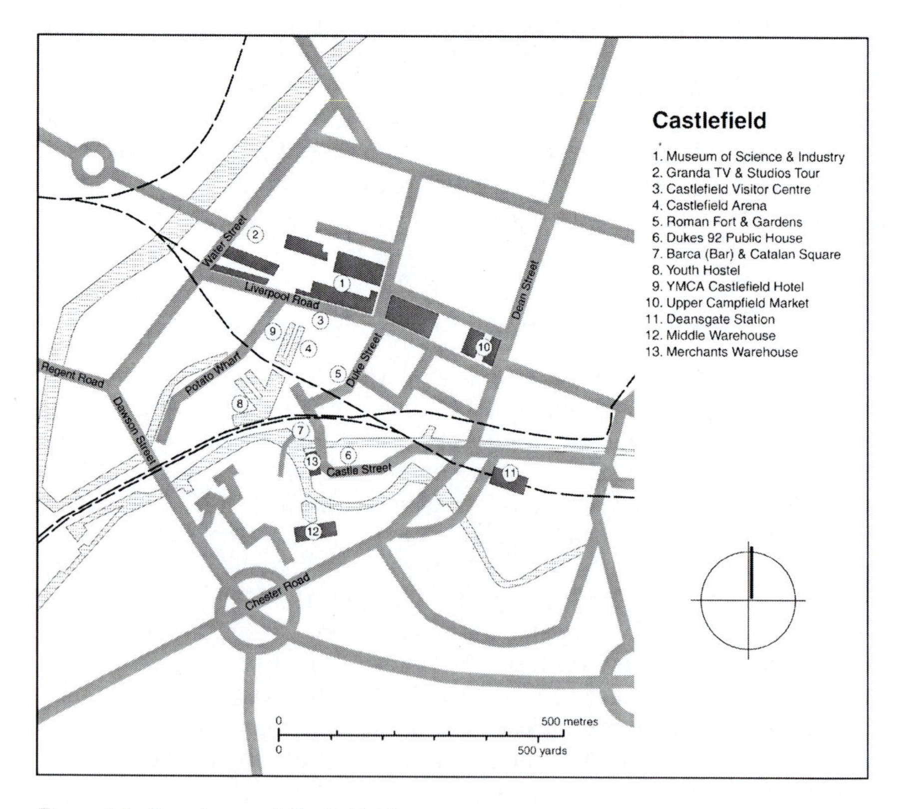

Figure 5.1 Street map of Castlefield.

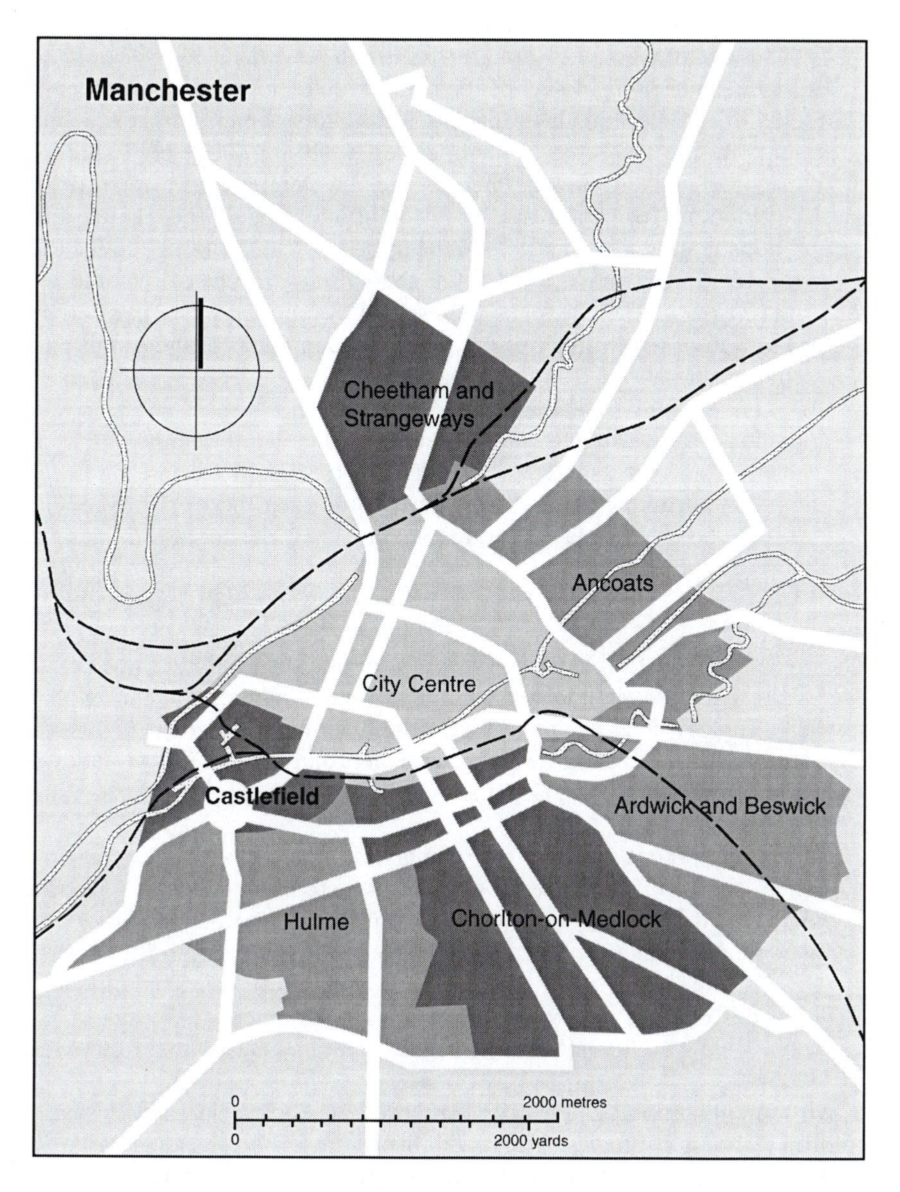

Figure 5.2 Context map of Castlefield.

preservation strategy for the existing architecture, and selected aspects of its industrial history offered a unifying and homogenizing theme for the area. Simultaneously a range of visitor attractions were put into place. In 1983 Liverpool Station was inaugurated as the Museum of Science and Industry, followed in 1986 by the opening of the G-Mex Exhibition Centre in the old Central Station. Granada Studios opened in 1980 and in 1983

Castlefield's Roman remains were recreated in a Roman Garden (Grigor 1995). These venues acted as signifiers for urban renewal.

However, the healing power of the reconstructed industrial past was not conceived as strong enough to dispel Castlefield's perception as a 'grimy back alley'. In the 1980s the council devised a vibrant public life based on the concept of 'mixed use', defined as the development of a mixture of tourist attractions, residential dwellings and office space for the creative and design industries. Leisure spaces started to be consciously designed for the area, representing a move away from the initial concept of an outdoor museum (Grigor 1995). The focus on leisure development was supported by the promotion of two historical features: the opening up of the canals and the recovery of Castlefield as a place for events:

> A walk along the area towpaths is all that is needed to see the potential it has for an increasingly popular leisure time activity, and the occasional colourful canal barge is already to be seen there . . . Castlefield will put the city firmly on the map as a tourist destination in its own right.
>
> (*Manchester Evening News*, 19 November 1982)

The mix of an industrial heritage landscape and leisure activities was envisaged to provide Castlefield with a competitive advantage on the 'global catwalk'. This is a clear example of how the recoding of place through heritage leads to the lucrative development of property. Culture provides a veneer for economic growth (Zukin 1995):

> The area's potential is one that retains its unique links with the past and keeps its essential character, but opens it up to more public activity and improves the environment to the extent that it becomes an attractive area for living as well as working and leisure – and indeed tourism . . . there's the opportunity to achieve something which is uniquely Manchester and which will be a world-wide attraction.
>
> (Manchester City Planning Department 1980: 14)

'Environmental improvement programs' proved to be crucial for these early cultural recoding strategies: 'The credibility of the Urban Heritage Park will be dependent on the quality of the environment within which various locations are located' (Officers Working Party 1982: page number omitted). The council's view was that the public perception of Castlefield could only be changed through a radical makeover and facelift for the landscape by stripping the place of its sensory coat of decay. Only when a uniform sensory transformation had taken place in the form of 'planting schemes (to enhance pavement areas), floor-scape treatment, cleaning of buildings, facelift schemes, painting of bridges and the relocation of unsightly activities, and street-furniture' would 'a cohesion and unity for the park' (Officers Working

Party 1982) be secured. Hence, businesses defined as 'unsightly activities' not fitting into the designed character of the Urban Heritage Park, or which might remind the visitor of the functional past of the area such as scrap-metal merchants, storage areas and car repair businesses, were relocated. We can see here how particular aesthetization practices, based on visual uniformity, are legitimated through the potential attraction of tourism. The way Castlefield is framed as a tourist destination establishes a hierarchical relation of the senses in which the visual dominates the other senses, which are muted or erased in order to support the visual seduction of the place. The 'ethics of engagement' are reduced to a passive visual consumption, a 'reduction and trivialization of the city as a stage of life' (Sennett 1990: xii).

The Central Manchester Development Corporation period, 1988–96

The 1980s riots in British inner cities such as London, Liverpool, Leeds, Birmingham and Manchester incited the Thatcher government to reassess life in British cities (Jones and Lansley 1995; Imrie and Raco 2003). This led in 1988 to the Action for Cities Programme, which involved a sharp reduction in the role of local government in policy-making. Urban policy started to shift increasingly from a welfare agenda to a business-growth agenda when the Conservative Party recognized business enterprise as the key for Britain's economic recovery. The problem of British society, as identified by Thatcher, was its 'dependency culture' which in her view had been fostered by previous Labour governments. It was assumed that a private enterprise approach to cities would have a 'trickle-down' effect and that the private sector would be able to provide the moral and economic regeneration of cities. Manchester's city council 'was forced to abandon its largely rhetorical commitment to 'municipal socialism' and to embrace a 'new realism'' (Williams 2003: 60). The foundations for an entrepreneurial politics were reflected in the change of the city's slogan from 'Defending jobs, improving services' to 'Making it happen' (Cochrane *et al.* 1996; Williams 2003).

The agencies to promote this change in derelict industrial sites were Urban Development Corporations (UDC). Centrally funded and controlled by the Department of the Environment, UDCs' main aim was to promote urban renewal through private investment, the most famous example being the development of Canary Wharf in London's Docklands,[15] and with a further ten UDCs in the rest of Britain. One can identify a distinct UDC regeneration method based on the identification of flagship schemes, the drawing-up of development frameworks, the programming of land reclamation and transport development and the engagement in extensive place marketing (Centre for Local Economic Strategies 1992). The aims of the UDCs were mainly about physical regeneration to attract property investment and image-building to promote the consumption of place, while local social and economic issues were not addressed (ODPM 1998).

In 1988 the Central Manchester Development Corporation (CMDC) was appointed to develop Manchester's southern city centre fringe and to support 'culturally significant developments, in stimulating residential development and a major programme of environmental enhancement, all aimed at improving the quality of the built environment and at providing a framework for establishing the city centre's cultural distinctiveness' (Williams 2003: 247; see also CMDC 1996). As O'Connor and Wynne argue, the promotion of Manchester as a site for cultural consumption was purely economic: '[t]he CMDC . . . became a cultural intermediary cast in the role of justifying cultural value via its direct relationship to economic value' (1996: 61). With Castlefield, the CMDC inherited an area which had already received £7 million of public investment and was starting to move away from its decadent past. However, it had not succeeded in attracting any national or international private investment because 'of its poor environment and the very high costs of treating its many protected buildings and derelict and contaminated sites. The primary reason for setting up the CMDC was to remove or reduce these barriers' (Grigor 1995: 63).

Alongside a focus on urban heritage the CMDC aimed to follow the council's initial strategies by developing a strong tourism base in Manchester.[16] Castlefield was seen as a key factor in converting Manchester into a major European city by promoting an open-air café culture in the area. The first café-bar, Dukes 92, opened in 1991 in a Georgian stable block; in 1996 two railway arches under the Southern Iron Viaduct were converted into 'one of the most fashionable bars in Manchester': Barça (Hands and Parker 2000); in the late 1990s Quay Bar, Jackson Wharf and Nowhere Bar followed, attracted by the success of the area. The European lifestyle theme, much influenced by Barcelona as a role model, and evoked through the water features and a mixture of industrial architecture and modern design, would become a crucial feature in Castlefield's reinvented public life. As a CMDC board member states:

> Barcelona is a good example of regeneration and street-life . . . There are strong similarities between Barcelona and Manchester. In fact Barcelona is recognised as the Manchester of Spain . . . there is also the connection of the Olympic Games . . . we both faced similar problems: Barcelona turned its back to the waterfront, the same applied to the waterways in Manchester. Barcelona is an example of good practice.
>
> (Mr G, CMDC)

A series of informal visits and relationships were therefore arranged between the two councils, as Manchester was planning to encourage a 24-hour city and was bidding (unsuccessfully) for the Olympic Games in 1996 (and did so again in 2000). The outcome was a Catalan Art Festival in Castlefield in 1995 and the creation of a public art work, the sun-seat Mediterrània which celebrates, in the Catalan artist's words, the common pillars of both cities: sun, water and industry. A Mediterranean lifestyle theme informs the surrounding Catalan Square which is bordered by the Barça bar:

> Catalan Square is now a thriving, lively centre of Manchester's social scene – with an especially Spanish flavour. Nestled under some of Britain's oldest railway arches and surrounded by canals, the centrepiece of the Square is the trendy new cafe-bar BARÇA, with its glass frontage overlooking the terrace – and on the terrace, the largest seat in Manchester – the golden circular sun. Throughout the summer, hundreds of outdoor pleasure seekers can be seen every evening in the Square, draped on and around the sun of Mediterranean [sic], glasses in hand, meeting friends, watching buskers and soaking up the atmosphere. The area is completely distinctive – as new bars have to be in this city, the competition is so fierce! Sitting on the sun, people watching has become one of the city's key attractions – and will stay so!
>
> (Tucker 1997, page number omitted)

The quote illustrates how convivial sensescapes such as glasses in hand (tastescape), meeting friends (touchscape), watching buskers and people (visual and audioscape), are invoked for the reinvention of Castlefield's new identity. The new public life is based on a combination of sociability afforded through private commercial premises and the physical environment, which creates the ambience that the visitor consumes amongst a like-minded public. Notions of public and private merge as cafés and restaurants expand into public space and private interests shape the activities in these quasi-public spaces (Crilley 1993; Zukin 1995).

With regard to its marketing strategies, the city's political elite very early sought 'to align Manchester not with its competitors but instead with its role models' (Ward 2000: 1099). In Quilley's (2000) view one can talk about a 'script' that those involved in the regeneration of Manchester followed to successfully sell and fund the new image of the city, where 'the copy goes something like: Manchester is a major European city, it has cosmopolitan qualities' (Quilley 2000: 609). This is clearly reflected in the city's marketing strategies since 2000 (see Figure 5.3). It is important to acknowledge, as pointed out by O'Connor and Wynne (1996), that Manchester's promotion as a 'European city' is not just a calculated move to expand service industries or a marketing device. Rather it is also about asserting a distinctive identity:

> 'Europeanness' was a means of redefining the relationship of Northernness to Englishness in a way that by-passed London's cultural dominance . . . The 'European city' was about the possibility of new images of self, of the possibilities of transformation as Manchester attempted to map itself onto a transnational 'cosmopolitan' space.
>
> (O'Connor and Wynne 1996: 69)

Figure 5.3 'This is not Amsterdam, this is Manchester', Piccadilly Gardens, 2001. (Photograph by author.)

Castlefield's transformed urban landscape and reinvented public life were crucial in supporting this discourse and imagery. Simultaneously Manchester had one of the highest concentrations of unemployment, crime, poor health and deficient housing in Britain (ODPM 1998).

The CMDC was one of the urban development companies to spend most on improvements on physical appearance and, in this regard, Castlefield was considered a major achievement (ODPM 1998). By the mid-1990s Castlefield was a successfully landscaped Urban Heritage Park criss-crossed by sparkling canals, with a flagship public square: the Outdoor Event Arena, a vast 3500 square metre stage for events and performances inaugurated in 1993. In the same year Castlefield Management Centre was formed to take over the general maintenance of the area from the CMDC through its Urban Rangers Service. It was located in the Castlefield Information Centre overlooking the basin. In the next few years a mixture of hotels, bars and residential homes opened around the canal basin such as the Castle Quay Warehouse (1992), the Eastgate Office complex (1992), the Castlefield Hotel (1995) and a youth hostel (1995) amongst others. The promotion of *leisure space* became a key strategy for Castlefield, with the aim of transforming an empty place, lacking in public life, into an active place that would be incorporated into the new uses of the city centre. Its success was reflected in a increasing number of visitors to the Castlefield Information

Centre: from 6516 in 1993 to 20,383 in 1996 (Castlefield Centre 1996). At the end of the CMDC administration in 1996 the Castlefield area was attracting over two million visitors a year and was receiving a steady amount of private investment, mainly from local entrepreneurs such as Jim Ramsbottom. These entrepreneurs played an influential role in the redesign of Castlefield, especially in keeping the unique feel of the area and renovating old warehouses into desirable residential developments, bars and business locations (see Quilley 2000; Ward 2000).

After 1996

In 1996 an IRA bomb ripped apart Manchester's city centre and led to a re-evaluation of the city structure. Since then a £1.2 billion root-and-branch redevelopment has taken place which has dramatically transformed the cityscape, opened up the city to the river Irwell and contributed to its building boom since 2000 (Quilley 2000). At this time Castlefield fell back within the jurisdiction of the Manchester Council and a plan to complete its regeneration was incorporated into a number of parallel planning strategies for the city centre, most importantly *Castlefield: Strategy and Action Plan to Complete Regeneration*; *City Development Guide* and *The Manchester Plan First Monitoring Report* (Manchester City Council 1997a, b, c; see Williams 2003 for a detailed analysis). The main priority of these plans was to promote inner-city living by enhancing the physical environment of the city centre. High-quality design was a major feature in the city's development guide and regarded as the key to promoting a sense of place and civic pride. Three strategies specifically concerned public space:

(a) sustainable development, that focuses on making cities into more attractive locations to live and work in;
(b) promotion of mixed uses, seen as an encouragement of vitality and diversity in the city, the guiding idea being the concept of the 'urban village';
(c) design.[17]

Emphasized in these new policies is the accessibility of public spaces as places for engagement, a move away from the city as an anonymous place, especially highlighted in street descriptions such as: 'Streets should be designed as places for people to meet. The street should be a public space which promotes socialization' (Manchester City Council 1997b: 26). The success of these public spaces is measured by the provision of the visual legibility of Manchester – its architecture and design – both in terms of attracting people to use the streets and squares and in terms of providing a strong civic identity. The enjoyment of public space is reduced in such discourses to an explicit focus on its felt visual appeal:

> What a city looks and feels like is a crucial part of its identity. If people are to enjoy places, they should be visually interesting. Manchester contains many remarkable buildings which give it a unique sense of identity . . . Landmarks help people to orient themselves and find their way around: vistas create visual links both within and outside the area and draw people along a street. The Guide encourages outdoor spaces which are stimulating yet remain comfortable and human in scale; spaces which at all times are clearly defined and serve a useful purpose.
>
> (Manchester City Council 1997b: 13)

It is worth considering some of the paradoxes in this quote in relation to the planning ideologies which underpin it. While places should foster visual interest and diversity, social diversity is feared and needs to be restrained. One could argue that while the visual senses play a determinant role, the other senses, especially touch and smell, are regarded as dangerous and need to be regulated; this is implied in the phrase 'stimulating yet comfortable'. Thus spontaneity, the unexpected – inherent features of public life – are regarded as negative and threatening to the overall sense of well-being. In these documents public life and spaces are regarded as products that can be manufactured and organized according to the visions of planners. As stated in the above quote, public space should be 'clearly defined and serve a useful purpose', which implies single-use public space and throws up questions about who determines what is a 'useful purpose'. The concept of 'mixed use' fits into Manchester's marketing of itself as a European café culture metropolis. It also notably serves to address issues of passive street surveillance – an increasingly common component of the British trend towards a 'fortress city' mentality. Civic pride mixed with a discourse of control of fear is used to defend the surveillance of public places, illustrating Zukin's point that 'the streets are both aestheticized and feared as a source of urban culture' (1995: 267).

At the beginning of the twenty-first century Manchester is celebrated in the national press for its 'stunning and unexpected urban renaissance' (Hetherington 2006), leading Sir Howard Bernstein, the chief executive of Manchester city council, to state: 'We are a brand now . . . Everybody now sees Manchester as a distinctive, commercial, world-class centre in its own right' (quoted in Hetherington 2006). Today Manchester is a city of 422,900 inhabitants (Greater Manchester has approximately 2.4 million inhabitants). Since 2001, there has been considerable residential development in Castlefield (an estimated 1500 one- or two-bedroom housing units have been built) that is encroaching on the open space that once characterized the neighbourhood. Castlefield has approximately 3000 residents: 'it is difficult to assess how many people actually live in these developments as a significant proportion are bought by property speculators who then choose to rent them or just leave them empty' (policy analyst, Manchester City Council, personal communication 5 June 2006). With average prices trebling

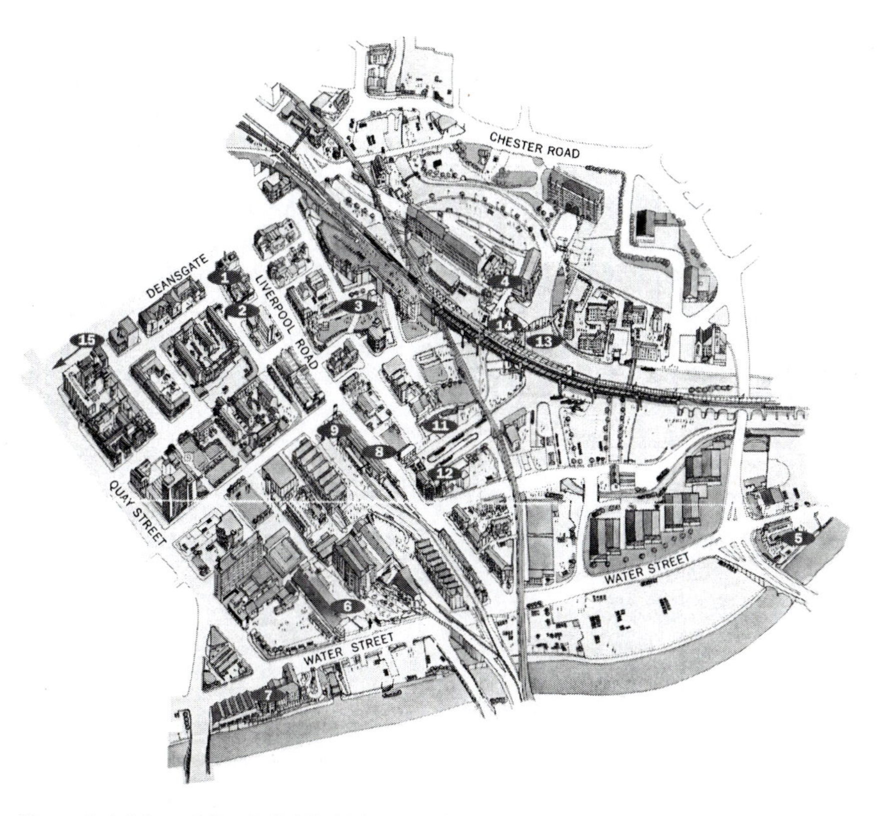

Figure 5.4 Map of Castlefield's Urban Heritage Park.
Source: Castlefield Management Company.

since 1997, and reaching £180,000 for a two-bedroom apartment in 2007, the area caters mainly for single, young, high-earning professional individuals (Figure 5.4). According to the 2001 census the area comprises 75 per cent one-person households and is constituted by a 95 per cent white population. Castlefield has become a flagship project for Manchester, and in 2004 it was awarded a prize for best practice in regeneration by the British Urban Regeneration Association, and a range of contemporary residence developments have received design awards. Inherent in most interviews with planners is a sense of 'Castlefield is what Manchester should be like.' From this review of the key policies that affected Castlefield we can conclude that the regeneration strategies have been successful in *manufacturing a new public space*. It remains to be examined what kind of public life has been created.

El Raval: from 'Barrio Chino' to cultural quarter

El Raval's twentieth-century planning history has been a story of constant threat of spatial restructuring. First, in 1875, Cerdá demolished parts of its north-east side to practise what later became his designs of the Eixample. Then, in 1934, the GATPAC[18] group, a rationalist circle of architects based on Le Corbusier's modernist school, drew up a project to demolish and rebuild large parts of the neighbourhood, a scheme that was stopped in 1936 by the Spanish Civil War. El Raval's poor housing stock stems from those times, as landlords abandoned the buildings designated to be bulldozed and left their inhabitants to their own devices (Figures 5.5 and 5.6).

Since then Barcelona's planning history has been shaped by two events that have had major repercussions on the redevelopment of El Raval. First, the democratization of Spain in 1976 and the accordance of regional autonomy to Catalunya in 1979 reinstated Barcelona once more as the Catalan capital. Second, in 1986 Barcelona was nominated to become the

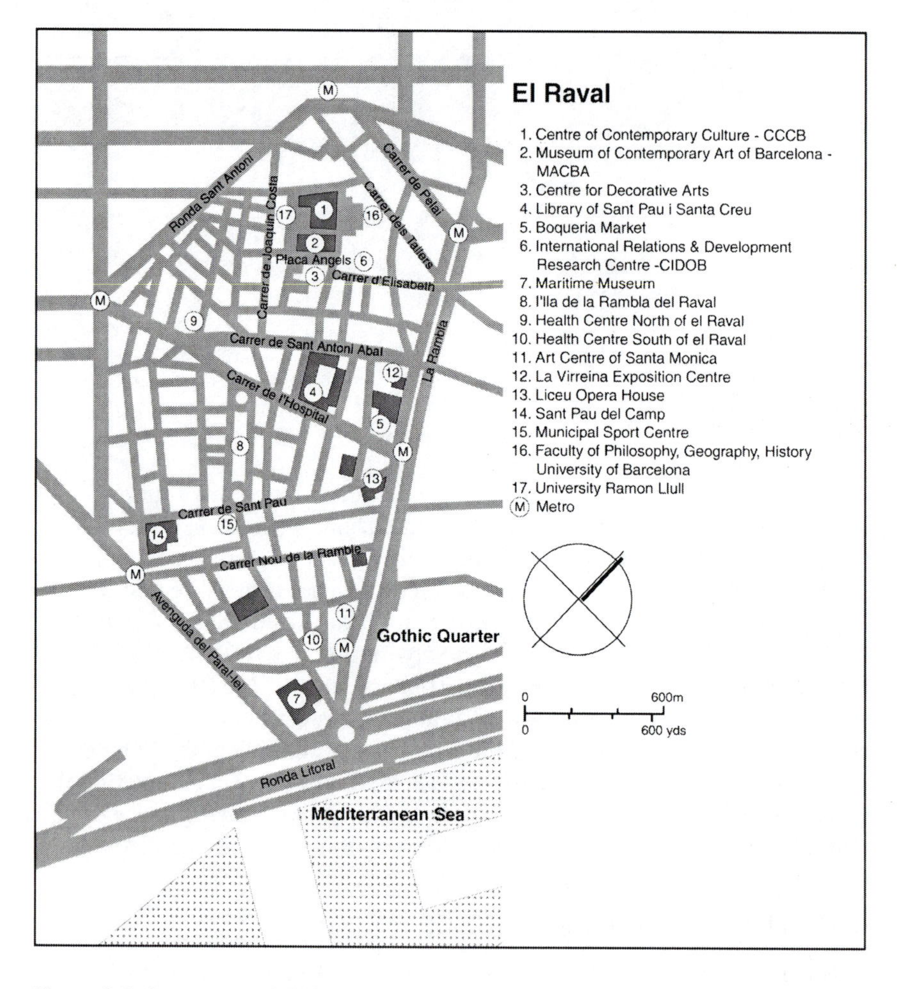

Figure 5.5 Street map of El Raval.

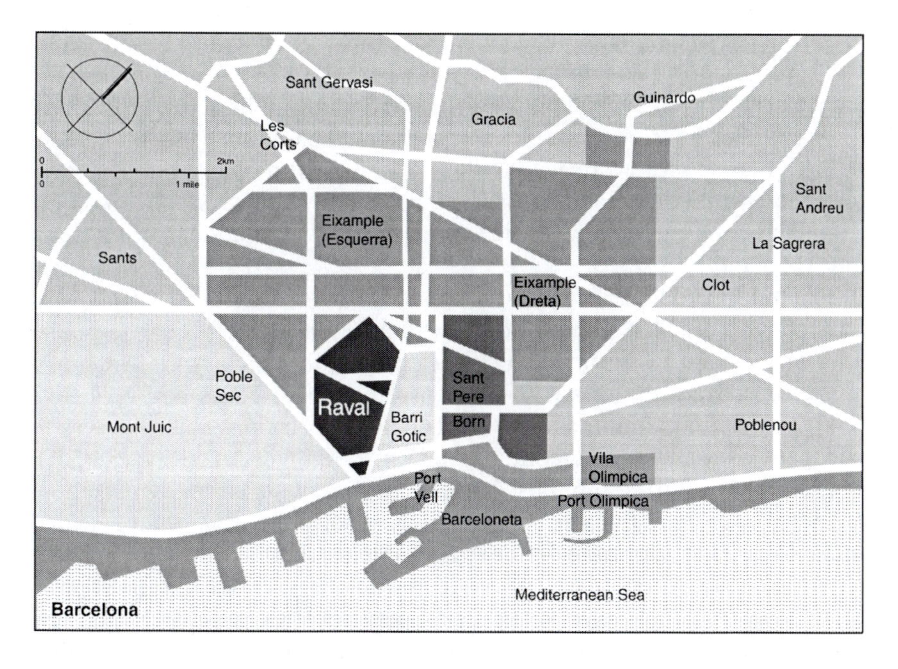

Figure 5.6 Context map of El Raval.

host city for the 1992 Olympic Games.[19] In 1977 the new democratic city council inherited a city that had been adversely affected by poor infrastructure, public institutions and civic facilities (Carreras i Verdaguer 1993). While Barcelona was certainly densely inhabited – it had grown from 533,000 inhabitants in 1900, to over a million in 1930 and 1.7 million in 1970 – much of its housing stock and public space was old and degraded. Hence, the first years of democracy between 1979 and 1985 were a period of reflection and urban reconstruction,[20] during which the city of Barcelona was subjected to a number of urban evaluation studies. In 1981 the first democratic planning department, 'Urban Projects Services', found itself with a disarticulated and incoherent city whose public spaces were colonized by cars. Those places that had not been the subject of commercial speculation were referred to negatively as 'empty spaces' (Ferrer and Caceres 1992; Garcia 1993).

Up until 1986 the 'Reconstruction of Barcelona' (as it was officially labelled) took place. It was a project limited by its financial resources and led by the architect Oriol Bohigas, who 'rather than devising some vast masterplan . . . concentrated on small urban interventions – new parks and public places – that would invigorate the city's various neighbourhoods . . . and weave together its outlying districts with its historic centre' (Bartolucci 1996: 64). It is generally agreed that this 'Urban Spaces Project' was characterized by an emphasis on design in public places, to resignify the city and provide an infrastructure that would support new communication technologies to ensure Barcelona's attractiveness to international businesses (Tello i Robira 1993). It is also important to emphasize the role that neighbourhood associations played during this phase in demanding public action to improve their city and guaranteeing the government's commitment to implement policies that supported residents' demands.[21] Creating more public spaces for collective use in areas such as El Raval has to be understood as coming from the strong civic ideals that informed the first years of democratic planning (McNeill 1999).

Planning in the Old City[22] was implemented through special plans of interior reform (Plan Especial de Reforma Interior, PERI).[23] The guidelines in these projects aimed to correct the original imbalance which had favoured private housing at the expense of public spaces and collective services. From an architectural perspective the principal objective was to 'recuperate the centre and monumentalize the periphery'. El Raval's PERI, conceptualized during the most extreme period of social marginalization, endorsed the idea that the transformation of the urban space would somehow lead to a solution of its social problems. At that time El Raval was one of Barcelona's poorest areas, riddled with drugs, crime and street prostitution and lacking basic social and health centres despite having the highest level of old and mentally ill people in the city. It was expected that its physical transformation would change Barcelona's perception of El Raval as a no-go area.

In 1985 the PERIs started to be implemented, and since then have been constantly readjusted according to the changing needs of the respective neighbourhoods. From the start El Raval's PERI had three main objectives. First was the creation of new public spaces in the neighbourhood with least green space in the city. This, rather than comprehensive slum clearance, was obtained by a method referred to as 'esponjamiento'. It meant that, similar to a sponge, small air-holes, or squares, would be cut into this compactly built neighbourhood. The second goal was to transform existing buildings into public institutions such as old people's homes, public libraries and social centres, amongst others. The third ambition was to build new public housing as well as encourage a general renovation or replacement[24] of old dwellings, which would lead to a better standard of sanitation in the district (Ajuntament de Barcelona 1983, Sust 1986).

Similar to the CMDC's aim in Castlefield, El Raval's regeneration has become increasingly culture-led. The target has been to resignify the neighbourhood for new activities by changing its urban structure and putting into place a range of cultural attractions, and by providing a pleasant, design-led environment that links El Raval with the rest of Barcelona's new design-conscious urban form. With the nomination in 1986 of Barcelona for the 1992 Olympic Games, the Catalan and Spanish governments, aware that the world's eyes would be set on the city, became involved in the rehabilitation of the Old City under a national planning policy called Area of Integrated

Rehabilitation (ARI).[25] Its objective was to improve and regenerate urban zones by coordinating public authority action and by promoting private initiatives. Through its management instrument, Promocio Ciutat Vella S.A. (PROCIVESA), a public–private company ordered the compulsory purchases, the relocation of families, the demolition of buildings and other tasks from 1988 (Abella 2004). In 2000 this limited company closed down and was replaced by Foment Ciutat Vella S.A.

The Olympic Games sparked off a symbolic and physical reassessment of the city's landscape in terms of its status and potential as a tourist attraction. The Olympic Games finally provided public and international investment to finance the city's large-scale public projects (Garcia 1993). Politicians and planners in the Old City realized that Barcelona's role as a tourist attraction had to be expanded to provide a new economy for the future. A crucial element in this image-building process became the campaign 'Barcelona posa't guapa' ('Barcelona get pretty') by the council which saw 6923 million pesetas (£27.7 million) spent on facelifting mostly the city's bourgeois *modernista* heritage (Balibrea 2001). While this can be read as an exercise in restoring civic pride, it also needs to be understood as a successful image-building exercise and political strategy 'which brilliantly exploits one communal, collective aspect of most of these buildings: the fact that their facades can be seen from the public space of the street' (Balibrea 2001: 191). Moreover, the way that the Olympic Games were marketed as the 'Catalan Games' ensured the public support of Barcelona's citizens: 'The Olympic Games generated a civil fraternity, materially embodied and reinforced in every architectural and urban project, which was perceived as required by the event' (Balibrea 2001: 198). For the sake of the Games, Barcelona's citizens stoically endured a number of highly contested destructions of favourite areas, such as the Barceloneta – an old area at the beachfront with 'chiringuitos' (picturesque wooden stalls selling fresh fish at very affordable prices) – which gave way to Barcelona's new designer waterfront. The moral cleansing of the Ramblas and the waterfront for the Games led to a re-evaluation of the Old City in terms of its use and symbolic role and clearly affected the policies implemented in El Raval.

Similar to place-making strategies in Castlefield, Oriol Bohigas, then Barcelona's head of urban planning, argued that history and heritage provided a collective identity for Barcelona's citizens, not to mention a marketable image. In his view a homogenized city planning approach was needed to support a unitary representation of the city based on heritage and the creation of public spaces (Bohigas 1986). There are many critical voices which have pointed out that this sanitization of the city's heritage was simply a means to satisfy global tourism (Heeren 2002; Balibrea 2001). The simplification of the city into one easily digestible theme, along with the conscious reduction of its sensory exuberance, which characterized the Mediterranean cityscape for centuries, creates a 'new homogenized urban ethnicity: the Barcelonity' (Delgado 1992: 19; Delgado 2005). Public space design has become a crucial marketing image of Barcelona, signifying a city that has successfully accomplished the move from industrial to post-industrial economy (Balibrea 2004). The first fractures between the different visions of how Barcelona's urban planning should evolve emerged at this time. The Old City's neighbourhood associations (which El Raval is part of) organized a campaign 'There is hunger here', which began in 1987, and drew attention to the poverty and social problems of the area. The campaign rallied the general public into demanding that the regeneration should tackle social problems rather than provide only a physical facelift to the area.

Regeneration in practice, 1990–2000

The PERI divided El Raval into two areas with different planning strategies. The northern part of El Raval, described as 'Del Liceo al Seminari' (From the Opera House to the Seminary), was based on cultural regeneration, and forms the main focus of this book. Southern Raval was subjugated to 'el Plan Central' (the Central Plan), a regeneration based on Franco's urban renewal plans that aimed to penetrate the neighbourhood with a broad avenue – a second Ramblas – and focused on enhancing commerce, fostering new economic activities, and providing the area with a solid social infrastructure through old people's homes, healthcare and sport centres (Clotet 1981; Ajuntament de Barcelona 1991, 1995). Regeneration started in the early 1990s at the northern and southern peripheries of the neighbourhood, working its way inwards and aiming thereby to affect the centre of El Raval where the toughest social problems and poorest housing stock were located. Simultaneously, the regeneration was expected to expand concentrically from the heart of the neighbourhood by creating a cultural quarter in the north and through the gradual bulldozing of El Raval's centre in order to create, in Haussmannian fashion, a boulevard into the southern neighbourhood (Figure 5.7).

In line with the aim of developing the symbolic economy of the city, the objective in northern Raval was to attract new cultural practices into an area packed with a large amount of so-called empty 'container buildings' (such as children's homes, a royal almshouse, hospitals or convents) by converting these into museums, cultural institutions, research centres and so on. The planning proposal envisaged additional public spaces such as parks and squares intended for 'flexible use', which would be open to both residents and outsiders. Cultural activities such as concerts, films and so on were expected to spill out from these cultural hotspots into the new public spaces. The Centre of Contemporary Culture of Barcelona (an exhibition space and research centre), housed in a former city orphanage, opened in 1994. The Museum of Contemporary Art of Barcelona (MACBA) designed by the North-American architect Richard Meier opened in 1995, and provided an internationally recognized architectural aesthetic. Reflecting Meier's unique style, the white-washed glass-fronted building would from now on be pictured in international travel and architecture magazines and attract a steady flow of visitors. Similar to the Paris Beaubourg area and the Centre Pompidou, the new museum was planned to serve as a catalyst for the neighbourhood's regeneration. An increasing number of culture-related institutions and businesses such as research centres, bookshops, a large number of art galleries, antique shops and designer shops settled in close proximity.

When analysing planning documents of El Raval's regeneration during the 1990s one is struck by the dominance of a medical discourse that describes this neighbourhood as an ill patient that needs to be operated upon: for example, the need for 'a surgery model that eliminates the non-recuperable' (Ferrer 1997: 68). This discourse operates through an emphasis on 'hygiene' to legitimate the demolition of large areas of housing mainly located in the south. As discussed in the previous chapter, the notion of purity or in this case 'health', in city planning is not new. It is frequently linked to the creation of a social order as social hierarchies are played out in space. What is shocking, however, is how these sensuous metaphors are naturalized as being intrinsic to the inhabitants of the area and are invoked in the name of urban liveability. The regeneration process is thereby portrayed as a positive transformation of perceived 'sick' or 'impure' social geography. The degraded zones and activities need to be 'cut out' as they are in danger of 'infecting' (sensuously) the newly regenerated zones. Economic and symbolic considerations involved in the resignification of property are disguised in a discourse of civic-mindedness. This is a sensuous expression of power: 'Renovation participates in the medicalization of power . . . This power is becoming more and more a "nursing power". It takes responsibility for the health of the social body and thus for its mental, biological or urban illnesses' (Certeau and Giard 1998: 139). A clear example of the power played out in this hygienic attitude is reflected in the demolition of whole blocks of houses containing 1384 dwellings and 293 commercial premises in order to construct the new Rambla of El Raval (*La Vanguardia*, 28 April 1998). The consequence was the disappearance of El Raval's 'hardest' streets, the focal spaces for prostitution and drug-dealing. It goes without saying that this was also the area where Raval's most vulnerable citizens lived.

So far, the strategies of cultural regeneration in the northern part of El Raval and the social regeneration in the southern part have exacerbated the already existing separation between these two areas. The south of El Raval now firmly lies in the shadow of its northern cultural quarter. The new Rambla del Raval has not yet attracted the expected number of visitors and so far has had little commercial success. At the time of writing another grand project is planned halfway up the new Rambla in order to inject economic revitalization. By 2008, a US-style mega-complex will have been built that includes a five-star hotel with an illuminated exterior, offices, commercial spaces, social housing and the Catalan Filmoteca.

Figure 5.7 Map of El Raval's regeneration, 2003.
Source: Promoció Ciutat Vella S.A.

While the regeneration started from a philosophy of restoration and aimed at 'recuperating' buildings and maintaining the 'character' of El Raval, the political will to implement this did not prevail (Font Arellano 1991; Claver-Lopez 1999; Garcia 1998; Heeren 2002). Consequently much of El Raval's housing has been demolished and its original spatial form has suffered extreme changes. Despite an explicit policy of providing people affected by the demolition with housing in a new building, this has largely not been the case. El Raval's main neighbourhood association was eager to proceed with the regeneration quickly and therefore only supported neighbours with legal claims. However, the most affected areas of demolition in southern Raval were characterised by people living without legal rent contracts who did not have the means to resist relocation or expulsion. This policy was increasingly criticized by planners and the media and led, by the end of the 1990s, to a second phase of the PERI in which emphasis was put on renovation rather than demolition.[26]

The consequence: 'Athens? Berlin? Milan? This weekend, El Raval!'

The negative reputation of El Raval as a red-light district has been dramatically transformed since the late 1990. A series of promotional campaigns by the council and PROCIVESA worked at modifying citizens' and residents' perceptions by promoting the cultural consumption of El Raval based on an eclectic mixture of history and new designer culture, captured by headlines such as: 'Athens? Berlin? Milan? This weekend, El Raval!' The main emphasis of the campaign lies in encouraging outsiders to explore through walks and open-house visits El Raval's unexpected and unknown historic richness '[h]idden most of the times underneath a thick cover of neglect, of dirt – also metaphoric – of flagrant degradation' (Gabancho 1991: 18).

One of the first actions by the city councils to stop undesirable activities was to issue a policy in 1992 to control the activities of the area: 'Plan de Usos' (Plan of Uses).[27] This was really a policy against brothels, lodging-houses and bars, and consequently a legal strategy to impose a particular spatial discipline. It is important to emphasize here the crucial social role that brothels, lodging-houses and similar businesses played in this neighbourhood. They took in and offered shelter to innumerable immigrants (increasingly illegal ones from outside the European Union). They also provided places where prostitution could take place. By destroying these 'undercover' spaces of reception, transitional residents and the illegal, non-registered population of El Raval was reduced and rendered more controllable by the police. Prostitution was either expelled to other parts of the city or forced on to the street. The next step in the regeneration process was the strategic replacement of a prostitution zone by a cultural and university site (Garcia 1998). Hence regional and national research centres and university faculties were moved into the area (one opened in 1996, the other one is expected to open in 2006). This is documented in this way: 'the addition of

Table 5.1 Overview of El Raval's and Castlefield's regeneration

Year	*Castlefield (187 ha)*	*El Raval (109 ha)*
Year	*Residents*	*Residents*
1996	Approximately 250 residents in Castlefield, 2000 residents in city centre	34,871 residents 3316 non-Spanish residents
2001	Approximately 1200 residents in Castlefield*	37,489 residents 8748 non-Spanish residents
2006	Approximately 3000 residents in Castlefield; 7154 residents in city centre	47,064 residents 21,165 non-Spanish residents
	History of regeneration	*History of regeneration*
1979	Designated Conservation Area	
1980	Granada Studios opens	
1982	Designated Urban Heritage Park	
1983	Museum of Science and Industry opens	
1985		Special Plan for Interior Reconstruction starts (PERI)
1986–98		Old City designated as an Area of Integrated Rehabilitation (ARI)
1988	1988–96 Regeneration by Central Manchester Development Corporation (CMDC)	1988–2000 Promocio Ciutat Vella S.A.
1992	Castlefield Management Company launched	
1993	Outdoor Event Arena built	
1994		Centre of Contemporary Culture of Barcelona (CCCB) opens
1995		Museum of Contemporary Art of Barcelona (MACBA) opens
1996		Ramon Llull University opens
2000		Rambla del Raval is inaugurated
2001		Foment Ciutat Vella S.A. takes over from Promocio Ciutat Vella S.A.
2002	Granada Studios and Castlefield Management Company close	
2006		University of Barcelona faculties of philosophy, geography, and history open
	*Investment 1988–1996***	*Investment 1988–2001*
	Public: £7million Manchester City Council; £101 million CMDC	Public: £750 million Barcelona City Council; European Union; Spanish Government
	Private: estimated £300 million +	Private: estimated £1.16 billion

Sources: *ODPM* (1998)
Office of National Statistics (2004)
*The census was undercounted by 37,000 people and this affected particularly the city centre (Manchester City Council, personal communication)
**After 1996 Castlefield fell back under the jurisdiction of Manchester Council and investment and expenditure in this area is subsumed under the City Centre ward.

Sources: Foment Ciutat Vella S.A. (2003)
Ajuntament de Barcelona (2007)

new users to this part of the city has neutralised the old marginal activities' (Ajuntament de Barcelona 1996).

Since the late 1990s another less-planned social process has deeply affected the socio-spatial development of El Raval, namely the unprecedented arrival of non-European immigrants from Pakistan, Morocco, the Philippines and North Africa. These new migrants have taken over many of the flats no longer wanted by Spaniards and have gradually transformed the sensuous geography of the streets. In 2007 El Raval was the most ethnically mixed neighbourhood of Barcelona, with 45.4 per cent non-Spanish residents (Ajuntament de Barcelona 2007). El Raval hosts more than 70 nationalities and has become one of the hippest neighbourhoods (*Time Out* 2004), packed with a mix of designer bars and restaurants, clubs and ethnic shops, and also attracting a large numbers of European migrants (Subirats and Rius 2005). Gradually more Barcelonians are venturing into El Raval, and the Museum of Contemporary Art has become the second most-visited museum in the city. While in 1999 it only attracted 150,000 visitors, in 2006 numbers reached 460,000 (*El Pais*, 23 July 2006). At the same time El Raval remains the poorest area of Barcelona. While employment has risen significantly, newly created jobs are mainly taken by outsiders to the district, given the low skills and education levels of residents in the Old City. Family income is clearly falling and the rising housing market is forcing local residents to move. The south of El Raval, however, still has the worst health indicators of the city (Subirats and Rius 2005).

My analysis of El Raval's regeneration strategies suggests that one can speak here of a *politics of substitution of public life.* All the policies applied have aimed to gradually expel unwanted marginal practices and social groups in order to attract new uses and activities to the area. However, as I show in the remaining chapters, this is not a straightforward or uncontested set of processes. In Table 5.1 opposite I briefly sum up the main information about each area in terms of residence, history of regeneration and investment.

Conclusion

My historical analysis clearly shows that these neighbourhoods did not lack public places before the regeneration but that a discourse of marginality and negative perception was constructed both in media and planning documents in order to support the radical physical restructuration of each neighbourhood (see also Lees 2003). This has led to an image of the neighbourhoods' future that promotes a new urban lifestyle associated with the regeneration and which implies that only regenerated areas are desirable areas – and which fails to acknowledge that the marginality of these areas is largely due to a consistent lack of financial investment or institutional support.

Reflecting the trends of entrepreneurial urban politics both regeneration strategies have been based upon public–private partnerships. This means that councils have provided the legal infrastructure to develop the areas and provided a first financial injection to instigate the regeneration. These regeneration strategies are distinct from earlier urban redevelopment schemes in their conscious focus on environmental improvement of public space that seeks to change the perception and image of the place and attract new socio-spatial practices, with a particular view to attracting private investment, which has been successfully achieved.

Both neighbourhoods resorted to similar culture-led regeneration strategies based on the integration of 'classic' cultural facilities such as museums or art galleries and the promotion of a cultural consumption of place through heritage trails, bars, restaurant and redesign of public space. Densely inhabited El Raval needed to change the perception of its public places by substituting one form of public life for another, and in the process expelling unwanted activities and inhabitants and attracting new ones. Castlefield, on the other hand, needed to consciously manufacture a new public life from scratch to bring people and activities into an empty area. Yet while Castlefield has become a flagship project for Manchester, symbolizing the European lifestyle and vibrancy that the city of Manchester wants to promote, El Raval is still a sore thumb in Barcelona. One could say that it is being 'fabricated' to fit into the overall image of Barcelona as a fashionable, forward-looking twenty-first-century European city.

Sensory paradigms are used to legitimate these spatial strategies. Regeneration discourses argue that a sensuous-physical transformation of public places can attract different spatial practices and generate a new public life in these neighbourhoods. By modifying the physical environment, authorities and planners are aiming to regulate the social relations that can arise in these new landscapes. Designed landscapes afford particular spatial configurations that affect how people engage with and in particular places. The questions that arise are: what kind of public life do official agents envisage? Can urban life be 'managed' in such ways? And how do 'socially embedded aesthetics' express themselves in the experiences and practices of everyday lived space? Such questions cannot be answered through the analysis of documents alone, but need to be assessed in the remaining chapters through Lefebvre's trialectic of the conceived, perceived and lived space.

6 Planning regeneration

> Streets aren't the same anymore [in El Raval]. When a narrow street becomes a wide street, the building is not the same anymore. When you take a building away, the one in front feels naked. Houses change, if you touch one bit it all moves.
>
> (architect R in El Raval)

Individuals are positioned in urban environments 'with different degrees of structurally generated power' (Herbert 2000: 555). This means that, for example, those building urban landscapes such as architects, politicians, urban planners or policy-makers, will experience and conceive this process differently from residents or visitors. Power in the built environment is framed through various socio-cultural dimensions and the manipulation or modification of sensescapes and spatial order is one of these. Whether it entails demolishing buildings or adding street-lights, every insertion in the urban landscape alters the sensuous geography of a place and rearranges experiences and associated meanings. In other words, the transformation of sensescapes redefines the cultural politics of place.

Meanings in the built environment are not fixed but are constantly open to contestation, subversion and appropriation. As Dovey (1999: 183–4) explains:

> [b]uildings necessarily both constrain and enable certain kinds of life and experience; they are inherently coercive in that they enforce limits to action. This coercion is a large part of what enables agency in everyday life to 'take place' . . . Enabling and constraining are poles of a dialectic of coercion in architecture. Architects 'manipulate' spatial behaviour – the issue is not whether but how they do so.

To understand the entanglements of power and the relationship between 'official' accounts of public space and 'local' experiences the next three chapters have been organized around Lefebvre's threefold distinction of the conceived, perceived and lived.

This chapter discusses the discourses by planners, politicians and architects that legitimate and realize the 'spatialization of regeneration'. I argue that the neighbourhoods' physical and aesthetic reorganization was conceived around the spatial techniques of 'accessibility' and 'designer heritage aesthetic'. It was based on a restricted view of the role public space should play and was expected to encourage predetermined types of public life. Supported by my own ethnographic observations the chapter unveils the sensuous ideological frameworks embedded in the regeneration process that inform the inclusion of commercially profitable experiences and practices in public space, and the exclusion of marginal ones or those that do not fit into the conceived vision of officials.

The spatialization of regeneration

We enter both neighbourhoods by leaving behind the busy traffic routes by which they are surrounded. Indeed, we have to physically cross a sensuous-spatial boundary that separates these areas from the rest of the city. In the case of El Raval we leave broad light avenues to walk along narrow winding streets bordered by tall grey buildings. Its spatial structure offers an ever-changing play of lights and shadows moving with the position of the sun. In the case of Castlefield its sensory difference from the city is not immediately perceived in its spatial organization but by the lack of activity on its streets. The hustle and bustle of the city diminishes, as we enter this quiet area.

Neither neighbourhood was bulldozed or regenerated from scratch, but in both cases their physical forms have been mutated and shaped to produce a distinctive public landscape. Interviews held with planners and city council representatives show that the spatial rearrangement and aesthetic recoding of both neighbourhoods was achieved through two main spatial techniques. The first refers to an insistence on accessibility, making areas that were perceived as 'no-go areas' physically and symbolically within reach of a general public. The second technique entails the resignification of the vernacular historic landscape through a 'designer heritage aesthetic' to attract outsiders.

Accessibility

The strategy of access for Castlefield implied restoring it as a desirable neighbourhood in Manchester's public imaginary. In the early days, Castlefield was not regenerated with a specific public use in mind. Rather, emphasis was given to creating public awareness by resignifying this derelict area as a historically valuable conservation area. When the CMDC took over in 1988, accessibility continued to be a key concern, but it was interpreted with a much stronger commercial emphasis. Castlefield was perceived before its regeneration as impenetrable by developers because scrap-yards and concrete plants had taken over, and old warehouses and shipyards spatially dominated the place. This discouraged outsiders from entering the neighbourhood both physically, because of negatively perceived sensescapes with 'Alsatian dogs prowling and barbed wire', and psychologically, because it was perceived that 'there was no public area at all there' (A, a member of the CMDC). Physical access was facilitated by improving roads and creating a new landscape that would entice people in: 'One of the first things we did was to make the area open to Manchester, and that involved putting new accesses in, and down by the Visitors' Centre there was a first access, a grand access, showing people the way in' (A, a member of the CMDC). To invite people to explore Castlefield the landscape had to be 'emptied out' and offer broader vistas of the magnificent warehouses. Furthermore, canals had to be hollowed out and expanded further into the area, so that they were easily visible from Liverpool Road (Figures 6.1 and 6.2). Public space was not the already existing physical infrastructure of public roads and streets but 'publicness' had to be created by assuring visual accessibility for outsiders: 'we wanted to try and open views of the area by removing hoardings, demolishing buildings' (A, a member of the CMDC).

Public space is conceived here as a space that has to be officially produced, designated and branded 'public property' by a landmark: the Castlefield Information Centre. The Centre was built on the main access road, Liverpool Street, and strategically overlooks the new basin. Once the pedestrian or car driver has reached the information centre, a white modernist cube, the new public places and canals lie below as places to be captured by sight. New paths and benches offer vantage points as part of a tourist trail for the transient visitor from which to visually consume the superimposed layers of Castlefield's history, such as the white Merchants Bridge (built 1996 by Whitby Bird and Partners) that imitates the curving viaducts crossing the canals (Figure 6.3).

Figure 6.1 Castlefield's Urban Heritage Park in the early 1980s.

Source: Manchester Forum, Nr 18, Summer 2000.

Figure 6.2 Castlefield's Urban Heritage Park in 2000.
Source: *Manchester Forum*, no. 18, Summer 2000.

Figure 6.3 Merchant's Bridge. (Photograph by author.)

As O'Connor and Wynne highlight, these are landscapes organized around middle-class values so that '[t]he new-old spaces of urbanity were not the ones of the productive communities but the middle class stroller who had the time and cultural knowledge with which to stroll through the landscape and absorb the vernacular as aesthetic' (1996: 57). Castlefield's physical transformation implied, in the eyes of officials, a social improvement of the area. Reflecting Neil Smith's (1996) discussion of the gentrification of Manhattan's Lower East Side, Castlefield developers 'portray themselves as civic minded heroes, pioneers taking a risk where no one else would venture, builders of a new populace' (Smith 1996: 23). This involved constructing discourses in which uses and perceptions of what existed before the regeneration had to be vilified as 'uncivic' and 'uncivilized' to support the radical changes of the place:

> If you look back 15 years there was no social atmosphere to the place. It was a very dodgy area, a lot of the people down here were essentially criminals . . . Now we have over 2 million visitors a year from all parts of Britain and throughout the world.
>
> (C, Castlefield Management Centre)

The inviting features of Castlefield as a tourist venue contrast with the restricted access it provides, due to its high rents and property prices, for those wishing to settle down in the neighbourhood. As a CMDC member suggests, Castlefield was conceived as 'the city asset, a national and international asset . . . it wasn't a place where you built council houses . . . because of its status' (A, a member of the CMDC). Thus regeneration officials have a specific urban form in mind and a particular vision of how the physical structure of Castlefield should reflect particular social values. Similar to Jacobs' (1996) discussion of the creation of Australia's cities during the time of British imperialism, Castlefield is understood in planners' quotes as a 'wild virgin land' that needs to be domesticated by its physical transformation. Such a colonial discourse draws on the idea that the place will be civilized and 'accessible for public use' once it is invaded and settled. Spatial power relations are exercised by carefully controlling those who settle in Castlefield through planning practices and the type of housing provided (or the lack of it). Furthermore, in tune with the colonial discourse which devalues the existing land until it is impregnated with the settler's values, the existing pre-redevelopment public spaces are not conceived as such by the planners. Rather spaces need to be symbolized as public by relabelling the derelict places of working-class labour such as factories, warehouses, or canals as places of cultural value and distinction through a thematic blanket of industrial heritage (Rojek 1995).

At the start of its regeneration, El Raval was, unlike Castlefield, a neighbourhood already well known in the city's public imaginary as the infamous red-light district. During interviews carried out for this study, urban planners, politicians and city council-run housing associations positively associated the change of the sensory physiognomy of the neighbourhood with the 'normalization' of people living in it. Here the strategy of access became a key spatial technique by which to replace residents and their everyday practices with new ones. This is supported by the conviction that El Raval's reputation was only to be improved when its public space was remodelled and its public life renewed.

> Why this insistence on public space? Because focusing on public space resolves two problems in a neighbourhood. Apart from having the narrowness of a historic city centre, [El Raval] has a second problem, namely, that it has had the function of receiving the 'residual activities' of the city for many years. The permeability, the facility of penetration by the exterior . . . by the rest of the city, the opening of the neighbourhood to the city was our principal worry.
>
> (D, El Raval councillor)

As in Castlefield, metaphors of colonization permeate the official discourses. Accessibility is linked to the idea of making the values of the new 'civilized' city penetrate El Raval: 'all regeneration ideas of the historic city since Cerda start from the basis of letting the values of the new city penetrate the values of the old city . . . The values of the new city are hygiene, order, salubrious conditions, hierarchies and the specialization of functions' (D, El Raval councillor). The councillor cites as desirable for El Raval the values of the new city: that is, modernist design values informed by an ideology of purity and order. He thereby implicitly dismisses the existing public space and life of El Raval in negative terms such as non-hygienic, disordered and unhealthy, with no hierarchies and a variety of functions: an uncontrollable space with no clear boundaries and limits. His description of the valuable features of public space corresponds with Sibley's notion of 'spatial purification' as a key feature in the organization of social space: 'The anatomy of the purified environment is an expression of the values associated with strong feelings of abjection, a heightened consciousness of difference and thus, a fear of mixing or the disintegration of boundaries' (Sibley 1995: 78).

As in Castlefield, the physical landscape serves here as a symbol of deviance in need for social control.[1] The underlying assumption is that a regenerated public space, meaning here a purified space, will improve the social behaviour of its residents. This feeds into a second line of argument, which revolves around the use of a masculinist metaphor of conquest: penetrating the place symbolically with the civilizing force of colonization. The public spaces of El Raval, populated by outsiders, will civilize the existing population with their new spatial practices and values. Jacobs (1996) points out that such language is not innocent but connects the discourse of colonization and the exercise of power; it implies a social categorization of the resident population as inferior, marginal and uncivilized. It serves to legitimize the domination or transformation of the space and portray the colonization as a natural logic and as a social necessity.

El Raval is discussed by many of the officials as a 'degenerated place' because of its overwhelming sensuous features such as the lack of public spaces and the 'darkness of the neighbourhood'. These physical features are regarded as promoting marginal activities:

> It was a place where the sun never shone, where there were no areas where children could play, and then of course they did other things . . . Current urban changes have a lot to do with the social improvement of the area. It also helps other groups to enter. Before [the regeneration] few people would have settled down in the Old City and the very moment we change conditions people come back.
>
> (Planner L)

Consequently, the spatial technology of accessibility operated in El Raval through two planning strategies. First, it involved lighting up and hollowing out the area. This was structurally achieved by demolishing blocks of housing to create new public spaces promoted in regeneration slogans such as: 'In El Raval the sun is rising' (Figure 6.4). As we can see in the leaflet in Figure 6.4 the powerful hands of planners and discourses of light and purity were still as powerful in the 1990s as during Haussmann's time.

Al Raval,
surt el sol

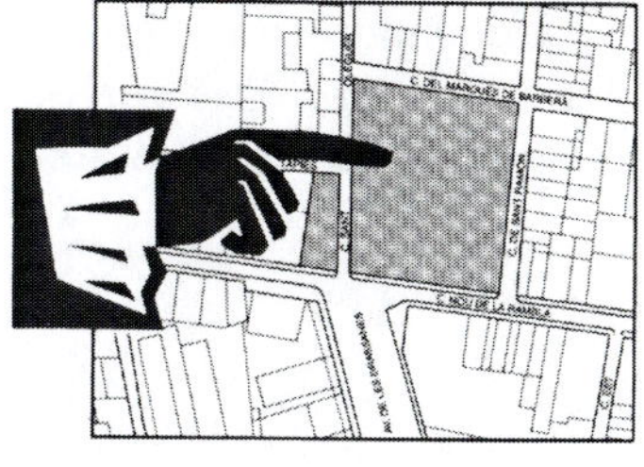

Millorem Ciutat Vella

El Raval ha guanyat 6.000 m² de sòl que donaran pas a un complex d'habitatges socials i dues zones verdes. Les famílies afectades han estat reinstal·lades a pocs metres, en vivendes noves al c/. de l'Om i la gent gran a la nova residència i apartaments per a avis al c/. Nou de la Rambla. Això ha estat possible gràcies a la participació de diversos Departaments i Àrees de la Generalitat de Catalunya i de l'Ajuntament de Barcelona i a la col·laboració dels veïns.

Figure 6.4 Promotional leaflet on El Raval's regeneration.

Source: Ajuntament de Barcelona and Promoció Ciutat Vella S.A.

Second, accessibility was based on attracting new activities that would permeate the area and would gradually dilute the existing public life and equalize its status to the rest of the city. In the view of a planner, permeating the area with new activities and residents would lead to its 'normalization': 'the fact that El Raval is becoming normalized also encourages people to get to know it, they lose fear and start entering its corners, its places' (E, spokesperson for PROCIVESA). Normalization is here related to 'designing out the fear' in these new public spaces by resorting to the planning ideologies discussed in Chapter 4. This involves first, dispelling impurity (letting the sun and air in) and, second, controlling social order by ensuring to Barcelona's citizens a socially homogeneous public space. In Davis's (1998) words this is the institutionally filtering-out of 'undesirables'.

Making the area 'attractive' also involved recoding the existing public spaces and surrounding buildings with new public amenities and cultural uses such as libraries, conference centres or exhibition spaces. The promotion of a 'cultural quarter' was meant to demystify negative perceptions, promote alternative uses of space and attract a new public. Needless to say, these public facilities were designed for El Raval's transient population of Barcelona's citizens and tourists rather than the neighbourhood's local residents, and were expected to lead to 'a social change in the uses of the neighbourhood' (Planner L). We can appreciate how culture is used here to shape power relations in urban development. As physical structures and uses of buildings are modified, the behaviour and spatial practices they support are expected to change and more marginal social groups and practices are likely to be driven out. This moral landscaping has major implications for the 'publicness' of public space when viewed as a dimension of political representation. In this way, its role as a public sphere is restricted and reduced. While I do not want to be too cynical about the social motives behind the regeneration of El Raval's landscape, it would also be wrong to ignore the commercial value that this central area of Barcelona has, once regenerated, for business, tourism and the property market.

Designer heritage aesthetic

The second important strategy identified in the physical transformation of regenerated areas is the resignification of the built environment through policy-led aesthetic practices. A common trait of Castlefield and El Raval is that their regenerated public spaces sensuously constitute a radical inversion of their past and nearby non-regenerated areas. Regenerated spaces create *public landmarks of difference* within the neighbourhoods – based upon and infiltrated by sensuous ideologies of order, purity and control. This becomes particularly clear when comparing the sensuous experience of the flagship areas, the cultural quarter around the Museum of Contemporary Art in El Raval, and the area surrounding the Urban Heritage Park in Castlefield. Formerly undesirable, cluttered, dark spaces – sensuously dense spaces – were redesigned into spacious, light, historic open-air museums by drawing on a common aesthetic code, which I define as a 'designer heritage aesthetic' (Figures 6.5 and 6.6).

The 'designer heritage aesthetic' consists of mostly pale shades: white, cream and grey colours predominate in both open spaces. The main squares

Figure 6.5 Plaça dels Angels around 1900.
Source: Arxiu Historic de la Ciutat, Barcelona.

Figure 6.6 Plaça dels Angels in 1999. (Photograph by author.)

in the flagship areas are characterized by newly designed elements interspersed with sandblasted historical features. In the case of Castlefield an amphitheatre covered by white canopies is set next to Victorian railway viaducts (Figure 6.7) and industrial warehouses. The whole is 'harmonized' within the aesthetic frame of the architect. The surrounding canal areas have been expanded to soften the hard industrial landscape of the space, and are regarded as important attractions to the neighbourhood: '[it] altered attitudes to the area . . . *that* actually created public spaces' (Planner K).

In El Raval part of a medieval convent and working-class housing had to give way to the white minimalist Museum of Contemporary Art of Barcelona. The surrounding vast square, lacking any seating or decoration, serves as a stage for the imposing building. In these regenerated neighbourhoods the already transformed is characterized by an absence of sensuous contrast. The tone of the main squares is subdued, as noise gets filtered through the spaciousness and is heard only as muffled background whispering. This restraint is all the more conspicuous in comparison to the activity on surrounding streets. Due to their scale, smells do not linger in these squares. Tactility is minimized by the smooth surfaces, no street furniture disrupts their uniformity, there are no boundaries, nothing to distract the senses from the self-referential celebration of place. A hierarchical relation of the senses is afforded by the spatial design of the environment in

Figure 6.7 Castlefield's Victorian railway viaducts. (Photograph by author.)

which the sensuous rhythms of the place clearly heighten the visual sense, whereas odours, sounds and tactile experiences are transformed into a supporting feature (Figures 6.8 and 6.9).

The sensuous values embedded within the 'designer heritage aesthetic' described so far are those of order and cleanliness, similar to Zukin's (1995)

Figure 6.8 Castlefield's Event Arena. (Photograph by author.)

Figure 6.9 Plaça dels Angels, bordered by museums. (Photograph by author.)

descriptions of public places in mid-town Manhattan where the motifs of local identity are chosen by commercial property owners and therefore lack traces of local life. In these spaces '[public culture] focuses on clean design, visible security, historic architectural features' (Zukin 1995: 36). One could easily argue that both squares and surrounding buildings could be anywhere and, in experiential terms, are interchangeable. They were conceived as prestige objects to attract investment and visitors to the place.

As cultural critics have pointed out, there has been a clear shift in contemporary cities towards producing a recognizable global iconography in which buildings by global architects have become status symbols of their post-industrial success and 'connote ideas such as cosmopolitanism, globalism and designer status' (Smith 2005: 413; see also Evans 2003). This standardization of sensuous and spatial organization makes these spaces recognizable environments for tourists. These are self-consciously designed spaces that produce familiar sensations and draw on a common cultural capital. They fit into Zukin's (1995) description of commercialized spaces for visual consumption. They are Lefebvre's 'dominated spaces': 'closed, sterilised, emptied out' (1991: 165). The main attraction afforded is the appearance of the place itself, and reflections in the water, in the glass walls of the museum, produce a narcissistic urban landscape, a celebration of its own monumentality. Beautiful in their own terms, distancing themselves sensuously from the rest of the neighbourhood, these regenerated landscapes were created as captivating landscapes to seduce the visitor at first sight with their visual qualities and formal arrangement, displaying what Edensor has termed 'visual imperialism' (1998). The 'visual imperialism' reaches its zenith at night when the surrounding locality disappears in the dark and only the heritage designer features are emphasized through elaborate light displays.

The building facades that were maintained after the regeneration both in El Raval and Castlefield were churches, warehouses, schools and convents (when they were not in the way of new developments): '[T]his scenographic arrangement is a compositional form which explicitly relies on a series of familiar, non-disturbing and comfortable views from [the] architectural past' (Boyer 1993: 121). Sights or buildings that might evoke awkward questions, reflect social inequalities or echo uncomfortable memories such as abattoirs, working-class housing, brothels or scrap-yards were demolished. Moreover, the most recent past – Castlefield's and El Raval's social and physical decay in the twentieth century – is brushed away, and not referred to.

This process is reflected throughout the regenerated areas of the neighbourhood. The absence of social conflict or complexity is deliberately planned into the sensuous spatialization of the landscape. Power is exercised through absence, by creating consensus, by deliberately covering over social inequalities and eliminating questionable features from the public sphere (Lukes 1974). Instead, the conserved shells of history are filled with new cultural functions: a Sunday school becomes a furniture design business, a chapel a recording studio and café; warehouse lofts convert into residences, a church into a bookshop; a charity house, once a lunatic asylum, becomes an art gallery.

Places are reinvented to conform to twenty-first-century sensibilities and economies. The present day's obsession with the past and sandblasted brickwork can be regarded as a reaction against the cosmopolitan and homogenizing impunities of modernism, reflecting, with its brick facades, vernacular architecture and national virtues (Samuel 1994; Urry 2002). In addition it is necessary for cities to show a forward-looking, entrepreneurial spirit and yet demonstrate controlled development. Hence, grand historic landmarks are recovered and blended with 1990s sleek architectural-designer landmarks, often produced by celebrity architects as '[c]ity images become essential in this marketing game: the kind of image that spatial pattern languages can foster and sell' (Boyer 1993: 125).

However, a 'designer heritage aesthetic' does not necessarily imply a sensuously homogenized environment. In fact, it is precisely the mixture of new design with local heritage that ensures the distinctive character of places, creating tensions between universal and particular features of place. In this sense we can speak of a 'vernacular postmodernism' (Urry 2002) where difference is enhanced to provide the necessary differential advantage on the global catwalk. Castlefield, for example, has two railway viaducts that span the canal basin and are still in use. Their imposing presence is supported by large castellated Victorian steel columns that dominate the area. The canals, the cleansed red bricks of the warehouses and the constant noise of rattling trains filling the Event Arena situate Castlefield as an industrial place, which was once one of the most important transport nodes in Britain. The square in El Raval is bordered on one side by non-regenerated eighteen-century working-class terraces. They are partly hidden by a Chillida mural so that their presence does not upset museum visitors (unless they look up). On the other side of the square is a sandblasted convent. Both borders refer to the deprivation in the area and its religious past. Thus in both neighbourhoods the vernacular features of place situate the embodied experience in particular gestures and understandings of place. The identity of each place is further enhanced by locally specific spatial practices, as I discuss in Chapter 8.

As the regeneration progresses the 'designer heritage aesthetic' permeates increasingly more areas of the neighbourhood. In El Raval new buildings in pale magnolia and discreet design contrast with older buildings marked by elaborate ornaments and encrusted dirt representing the passage of time (Figure 6.10). The new buildings contrast with old ones as they have no, or only shallow, balconies. In old Spanish streets balconies serve as thresholds between the private and public, where the inhabitants hang out their washing, decorate the edges with plants and flowers, and place birdcages. They are extensions of their living rooms. Conversations are held between neighbours and old people gaze silently on the life of the street as it passes by. However, these features are considered as the 'charm of poverty' (Planner Q). As the regeneration progresses balconies become, in the councillor's view, redundant:

> If we improve the housing conditions and ventilation, the outside, the street, loses its importance. The street was important for buildings and areas that did not provide a certain interior comfort . . . Those who live in new buildings do not have the need to consider the balcony as a way of relating, or as an extension of the street, because their houses are now comfortable.

Figure 6.10 New and old in El Raval. (Photograph by author.)

Yet, the erasure of balconies, the widening of streets has also an important consequence as residents turn inwards, away from the public life of the neighbourhood. The spotless new balcony-free and minimalist buildings and the new broad airy streets and public spaces of El Raval conform to the image and hard spatial boundaries of the modernist northern European city and are regarded as a desirable feature for a modernized Raval by official agents. Indeed, during the interviews, officials were eager to distance Barcelona from the image of a south European city and instead cited Swedish or German cities as role models of urban life.

The spatial technologies of accessibility and the reorganization of vernacular architecture into a 'designer heritage aesthetic' produce a new sensorial order of public space in El Raval and Castlefield. A striking feature for the pedestrian in these two areas is that they are strongly segregated into old marginal and new regenerated areas or, as I would term it, into *spaces of public presence and absence*. On a spatial level, this is linked to a zoning of these places where only regenerated areas are considered by officials as acceptable for 'public use'. They are made present in the public imaginary by enhancing them with signs and advertising them as tourist trails. These 'present spaces' are characterized by a complete transformation of their past sensuous landscape. They are marked by a subdued sensory experience and sensory rhythms coordinated around a visually harmonious landscape: an absence of strong odours, a lack of tactile difference, a coordination of textures and colour schemes, an erasure of traces of time in the sleek surfaces of historic buildings. These are sparse, carefully ordered public spaces – their silence is broken only by staged events (Figures 6.11 and 6.12).

As soon as the pedestrian leaves these predetermined routes, which are clearly signalled and guide her through the front-stages[2] of regeneration, she enters the back-stages or the *spaces of absence*. These are the spaces labelled as non-existent in the public imagination by official agents, as they have not been 'revalued' by the transforming powers of regeneration. Here, the haptic sense dominates, with dark colours, a variety of textures and surfaces, a lack of visual harmony, and a strong exposure to smells which confirm the way these places were before regeneration. In Castlefield, these are the canals still filled with rubbish, derelict factories and stretches of ruins and wasteland (see also Edensor 2005). In El Raval the back streets are a labyrinth of confined streets and alleyways. Balconies almost touch each other, washing hangs out from the windows, one can smell food being prepared, perceive the scent of washing powder and listen to the blaring noise of radios and televisions that make the pedestrian aware of a living city above. These spaces of absence are those that Bauman (2000) describes as 'empty or leftover spaces' where differences are made invisible by being prevented from being seen. These spaces are absent because they are rendered invisible by regeneration.

Figure 6.11 A traditional street in El Raval. (Photograph by author.)

Figure 6.12 A regenerated street in El Raval. (Photograph by author.)

> They are, we may say, the 'leftover' places which remain after the job of structuration has been performed on such spaces as really matter: they owe their ghostly presence to the lack of overlap between the elegance of structure and the messiness of the world (any world, also the purposefully designed world) notorious for its defiance of neat classifications.
>
> (Bauman 2000: 103)

Regeneration not only transforms the built environment, but as I discuss in the next section, in both areas the spatial techniques of regeneration manipulate spatial practices and serve to orchestrate particular expressions of public life in each neighbourhood.

Formalizing public life in Castlefield

Interviewees agreed that over the years Castlefield's open and public spaces have been pivotal in giving the area its unique character. Indeed its place identity has been based on differentiating itself from the rest of the city by promoting and securing its extensive open spaces which afford its competitive advantage: 'if we didn't have the open spaces a) we wouldn't have any performance areas and b) we wouldn't have the same sense of attraction' (C, Castlefield Management Centre). As the regeneration proceeded the organizing concept behind Castlefield's public spaces became leisure, with the view of creating a distinctive place identity. In the rhetoric of the interviewees, it served to promote the place for tourists and to attract businesses and luxury residences, all in order to generate economic rewards. Leisure had been conceived as 'passive leisure, so that people can walk around and see the canal infrastructure, the railway infrastructure, the older buildings' (A, a member of the CMDC). A paradox inherent in the 'leisurization' of regenerated neighbourhoods such as Castlefield and El Raval is that while leisure implies notions such as freedom and choice, in this context it becomes a highly circumscribed and controlled activity, limited to the particular possibilities offered in that place (Figure 6.13).

Leisure was planned around three main concepts that stimulated specific spatial practices geared towards 'consuming place' (Urry 1995). First, there was a heritage trail in which visitors are guided from one heritage site to the next by signs and a heritage map. Second, a 'Mediterranean lifestyle' theme was fostered, based on promoting café and styled bar enterprises in the basin. And, third, the canal area in the Urban Heritage Park was animated through staged events held in the new open spaces. Here, public life is envisaged as being a specific 'event' in time and place, leading to distinct commercial benefits:

> The events area was part of that overall strategy, making the area accessible, attractive, something to draw people into. And then off that framework you could attract developments to go on that framework, and we sought to attract leisure, retail, housing, and limited office development.
>
> (A, a member of the CMDC)

What was then the envisaged public life for Castlefield? Developers regarded public place as a residual space that had to be artificially animated in order to create an attractive experience for people. It was a contrived space – a link between commercial activities:

> [it] was always the intention that the outdoor spaces would really be a sort of link between the various operations within Castlefield, the spaces to hold activities or even just to go and sit in relative comfort away from the city centre in a more leisure based area.
>
> (C, Castlefield Management Centre)

But, the more successful Castlefield became in securing inward investment, and the more animated public space got, the more this strategy conflicted with an increase in private developments that reduced the available open

Figure 6.13 Visitors inspecting a map of Castlefield. (Photograph by author.)

space. Hence, the regeneration strategy had to ensure the increased *formalization of events* so that the public life in the area would not diminish despite the disappearance of open spaces:

> Very early on we recognised that we should formalize this activity, because all the events previously had been held in open spaces where buildings had been demolished and so we had a competition and winning designs came up with a scheme to put the events here.
>
> (A, a member of the CMDC)

This formalization of public life through events took place in both a physical and a symbolic form. On the one hand, a space was specifically bought by the CMDC to construct an Event Arena. On the other hand, the animation of Castlefield's public places was symbolically formalized through a series of regularly planned events in the Event Arena, organized by the Management Company, and the council-run bank holiday market. The latter drew on Castlefield's historic reputation as a market place, and stands sold a variety of bargain items, attracting large crowds from mainly lower-income groups from the surrounding areas. From its inception this was a controversial feature, colliding with the image that official agents had envisaged, as 'it brought in some people, the residents would think "undesirable" and there was conflict in that' (A, a member of the CMDC). Therefore, from initially being held monthly, the market became a event to be run only on summer bank holidays. Yet the market traders 'don't comply with the overall idea of quality regeneration in Castlefield. The quality of the markets has not come up with the rest of the area' (C, Castlefield Management Centre). From this quote one can conclude that it is not the market per se that devalues the area but the type of market, the consumption patterns it promotes, and the people it attracts that do not fit into the picture of an exclusive, 'quality' regeneration. Indeed, in recent years, the city council has invited more international stallholders onto the market selling French cheese, Italian olive oil, lavender and other continental produce and encouraging a more middle-class use of space. A specific, circumscribed notion of public life is promoted that revolves around particular socio-sensuous expectations and the orchestration of middle-class spatial practices.

An animation of the area that fits in more with the planners' vision of 'quality regeneration' and in their view attracts the right groups with disposable income were various themed events, such as a Latin festival sponsored by the San Miguel brewery, world music performances or a family carnival held in and around the Event Arena from April to September. The promotion of outdoor activities and the animation of public life by outdoor events were part of the Mediterranean theme that was envisaged for Castlefield and heavily marketed in its promotional materials (Figure 6.14).

These events, ranging from world music festivals to avant-garde theatre performances, are a desirable attraction and are purposefully designed to

Figure 6.14 An event held in the Event Arena. (Photograph by author.)

cater for a middle-class public. As a member of the Castlefield Management Centre explains:

> Essentially you cannot put on events for people who are not willing to spend a basic amount of money for a day out . . . There is no point in holding a big sponsored event to try to market a product to somebody who either can't or won't pay for it, so essentially it's going to be a middle class sort of activity profile.
>
> (C, Castlefield Management Centre)

This quote clearly shows how the exclusion of certain groups from Castlefield's public life was based on the conscious shaping of taste values, spatial practices and consumption patterns in the provision of events and activities. Economic considerations informed the forms of public life encouraged in the area. Social groups that lacked the disposable income were not targeted by these events. Ultimately, the economic interests of sponsors, bars and other leisure premises, while not directly involved in the organization of events, determined the uses of and experiences in a collective public space.

Castlefield's 'continental' public life reflected in sensuous affordances inscribed in its regenerated landscape was regarded by official voices as a further attraction of the area: 'Castlefield is, or can be, a pleasant place in the evening with the lights sparkling off the water and its bridges. It can be quite a romantic sort of place' (A, a member of the CMDC). This can be understood as a strategy to deliberately contrast the associated negative myths of the northern industrial city and was in tune with 1990s commercial demands to create a 24-hour city, as the next quote illustrates:

> That was the overall ethos at the beginning of the regeneration of Castlefield. We could be a flagship area in this idea of continentalizing Manchester and making us a genuine European city with all the activities and trappings that go with that: al fresco eating, drinking, 24 hour operation, regular festival and carnival type activities.
>
> (C, Manchester Management Centre)

Public life gets reduced here to specific consumption practices that only cater for particular groups in the city: the young and those with expendable income. The continental lifestyle theme fits into the picture of the 24-hour city simplified into the 24-hour consumption city, in other words an 'economic continentalization' of the consumption experience. The drawbacks of this artificial creation of a continental lifestyle solely based on bars quickly became evident. The over-concentration of bars in one part of the city started to attract mainly large groups of men who did not conform to passive leisure practices and disrupted the envisaged public life, leading Castlefield to gain the reputation of being a 'drinking' area.

In Castlefield the diverse strategies for controlling and artificially formalizing a certain type of public life has led to a clear segregation and zoning of activities. Officials claim that this is a planned and positive outcome, as Castlefield is an area deliberately fragmented into different public uses: 'It's the fact that you've got all these different things, the history, the structures, the water, the bars, the market, it's just appealing to different people for different purposes' (Planner K).

Diluting public life in El Raval

The envisaged transformation of El Raval's public life, influenced by its historical and local particularities, was based on different criteria from that of Castlefield. Nevertheless, the comparison of these two neighbourhoods shows that parallels can be drawn in the sensuous ideologies that informed the redesign of these areas. In El Raval the regeneration of public space, rather than a tool to differentiate the neighbourhood, was conceived of as an instrument to homogenize El Raval with the rest of the city, both socially and psychologically. While physical homogenization was rejected, as it is the physical features which give the neighbourhood its unique marketable identity, El Raval's population and activities had to be amalgamated with the rest of the city by fostering a 'social mix'. This was desirable because 'when you have a specialized activity is when ghettos, good or bad ones, are produced' (Planner M).

The creation of new public spaces was regarded as a crucial element to improve its public life and to provide the neighbourhood and its inhabitants with a new public 'service', as 'public space has almost the character of a public institution, because of the activities and enjoyment that occur in it' (Planner P). Interviewees agreed that each new public square had to be purposefully designed to attract specific activities, to provide for a particular social group. Each new public space in El Raval was conceived to play a unique function within the neighbourhood. My findings suggest a notion of public life that leaves little space to chance. In fact, design is conceived here as a tool of control for determining how and by whom these spaces will be used. A striking mechanism to control the public use of most new public spaces, especially those surrounding the new museum, was their fortification with fences that closed at night.

Based on an artificial inversion of 'public space', planners claimed the right to shape access by portraying themselves as rescuing public space in the name of public interest. They argued that the social conditions, in particular the criminal nature of the area, demanded such drastic measures: 'There has to be some urban control, citizen security, salubrious conditions and cleanliness and you can't leave public space in the hands of some public space terrorists' (Planner P). Simultaneously, the creation of new public places was viewed as enhancing the attractiveness of the neighbourhood to outsiders. For example, to encourage a more diverse use of El Raval's streets, since the late 1990s a range of walks has been offered through the neighbourhood in order to educate Barcelona's residents about its historic value. The transformation of public space is crucial to these walks, as 'by recognizing the value of these spaces and creating interconnections between them, public space gains value and creates an urban itinerary' (Planner P). The description of public spaces creating an 'urban itinerary' is interesting, as El Raval is historically known for providing urban itineraries that made it easy for people to hide. 'Urban itinerary' in the present context does not refer to an unexpected, explorable or surprising space, but instead confers a highly controlled notion of public space in which people are guided along demarcated routes.

Two linked strategies are pursued to create a 'social mix' in the neighbourhood. First, the 'dilution' of the existing population and its marginal activities. This is achieved by a second strategy, namely, the colonization of public space by a particular set of new residents. What was striking in many of the interviews was the underlying discourse of 'natural displacement' of the existing population. As a member of the urban planning department summarizes: '[The criminal activities] are happening because one type of people occupy [El Raval] . . . When they start to mix they will be forced to disappear. Because people won't allow somebody to shoot up [heroin] in their door entrance' (Planner L).

The mere presence of 'civilized' newcomers is expected to eventually domesticate the public life of the area by forcing marginal activities to disappear, not by finding solutions to them but by moving them out of sight, drawing attention to how practices to impose 'civility' can be underlined by exclusionary meanings.

While historically El Raval was the first place where immigrants, first from southern Spain and in more recent years from Morocco, Pakistan and India, settled, the newcomers anticipated by planners and politicians to provide a 'social mix' were none of these. Inferred in the above quote is the attraction of middle-class Spanish/Catalan residents into the neighbourhood. These are expected to displace marginal activities and 'retake control of the political and cultural economies as well as the geography of the largest cities' (Smith 2002: 445). The discourse of 'social mix' is hence given a particular meaning, close to Neil Smith's concept of 'revanchist urbanism' which 'embodies a revengeful and reactionary viciousness against various populations accused of "stealing" the city from the white upper classes' (Smith 1996: 18).

The argument of 'social mix' is further used to support the entrance of private developers into the area. Not surprisingly, since the start of the regeneration in the early 1990s, the average price of a square metre of real estate in El Raval had risen by over 200 per cent in 2000, and up to 400 per cent in 2007. This has led to an inevitable and irreversible expulsion of socially vulnerable people, and also of children of long-term resident families who have had to move out of the area as they can no longer afford the increasing rents (Claver-Lopez 1999). This expulsion of people is conceived by some official voices as a desirable by-product of the regeneration. Hence the expulsion of people through a regeneration process is defended as the only way to improve the social life of the neighbourhood. To achieve a 'normalized neighbourhood' marginality has to be divided into smaller sections or, as a planner explains, 'diluted' in order to be 'treated'. El Raval's councillor at the time states: 'it was certain that the regeneration would lead to the expulsion of an amount of marginality. And also in many cases to the solution of it' (D, Raval councillor).

The displacement and reallocation of residents in El Raval has not occurred without contestation and pain, as I elaborate in Chapter 8. What took most of the local population by surprise and led to important housing problems was the incredible 'speed of the improvement of the neighbourhood' (F, social services). Most help provided by social services was in regard to the support of rent payments and the provision of legal advice to sometimes illiterate, and most of the time socially vulnerable, people, who could easily be taken advantage of by estate agents and easily coerced by 'people in suits' into moving with little compensation. This led a member of the social services to state: 'I often say cynically: we hold the population together while [planners and politicians] make the urban transformations' (F, social services).

Public support for the drastic changes in the neighbourhood was assured by organizing a grand celebration for each newly gained public space, such as the opening of the Rambla del Raval in 2000, the boulevard that now divides the neighbourhood. The start of construction work was celebrated with a number of events and a free lunch for residents. On the bare round square, dotted by palm trees, the mayor gave an inaugural talk, shaped by the discourse of 'civic recovery' and interspersed with slogans such as: 'In this neighbourhood we've got rid of thieves! Decent people live in this neighbourhood! El Raval belongs to its people!' Surrounding him stood the buildings to be demolished in the forthcoming months, and from their balconies mainly old people were silently watching the spectacle. Alongside the buildings big flyers advertised: 'El Raval of the people' and posters showed the new Rambla, proclaiming 'The Old City going to Europe: the new Raval'. Civic consensus is achieved by portraying the changes in El Raval as inevitable and part and parcel of Barcelona's new role as a European capital. As Smith (2002) highlights, urban regeneration is incorporated into neo-liberal visions of the city and portrayed as an inevitable consequence of a new global reality rather than a financial consideration.

When enquiring in more detail about the expulsion of the economically weak and transient population of illegal residents (there are no statistics, as many of the evicted were not registered in the neighbourhood), politicians responded by stating that the aim is to attract 'a new type of economically weak and transitory population' (D, Raval councillor), namely young couples and students. Ironically the biggest new social group attracted to

El Raval, non-European immigrants, was not mentioned during my interviews in 1998–9 or was discarded as a passing problem. El Raval and the historic city centre as a whole was envisaged as a 'learning ground' in which the daily life typical of a dense, marginal neighbourhood could be turned into a marketable consumption experience. The social problems that are particular to the social set-up of El Raval are transformed into a playground for developing life skills for tourists and the young middle class: 'You have to keep the historic city centre because it offers an urban diversity which teaches you skills for life' (D, Raval councillor).

When asked whether these strategies did not imply a clear 'gentrification' of El Raval, planners and politicians alike were eager to dispel this notion from the regeneration process. As Smith (2000) remarks in an article on El Raval in Barcelona's main paper, *La Vanguardia*, 'gentrification' is a word that has been banned from many recent urban planning schemes. Indeed, Smith argues that the notion of 'gentrification' has been replaced by the metaphor of urban regeneration and this is 'anaesthetizing our critical understanding of gentrification' (Smith 2002: 446). In the case of El Raval's regeneration, 'social mix' is the euphemism that has served to support the gentrification and physical renewal of the area. As we have seen, under the heading of 'social mix', some residents have been reallocated to different areas of Barcelona, others have been expelled. The architectural desirability of El Raval has been enhanced through policy-led changes in aesthetic codes and spatial practices aimed at an international middle-class public with the implications that the existing public life gets replaced.[3]

By 2000, the cultural quarter was becoming, in the official agents' discourse, a 'symbol' for the successful transformation of a neighbourhood that integrates new and old practices and becomes more like to the rest of the city:

> The council isn't interested in the MACBA, CCCB ... to be a Pompidou Centre for outsiders and to be talked about. We are much more interested in the MACBA being a factor, an instrument to normalize the area and make it attractive for all kind of uses. In those terms I think it is working quite well; maybe it is not a great centre of attraction for tourists, but that isn't needed. The objective is the territorial balance of activities and I think that is happening perfectly. Moreover, it is becoming a bit like the symbol of transformation which is good.
>
> (E, spokesperson for PROCIVESA)

A visible indicator for El Raval's councillor that cultural regeneration is reinvigorating the public life of the neighbourhood was the fact that empty retail facilities were being filled again by 'specialized retail . . . linked to culture, to the university, to immigration . . . if streets recuperate their commercial capacity it will be because they find a specialisation that makes them attractive'. Off Hospital Street a whole street is devoted to second-hand stores for young people; nearby another street is attracting record shops and clubwear boutiques. Indeed, the councillor of El Raval made a careful distinction between what is regarded as negative economic gentrification and a desirable 'cultural gentrification':

> It is generally young artists, young designers [that are settling in]. It is not an economic gentrification. They are people that are taking over the centre of the city because in the city centre they find the values that are linked to their life values, not because the value of the centre has gone up. This is an absolutely desirable gentrification because it is a social group that finds its values in the centre. Not because the powerful social groups occupy the centre, but because the centre has recuperated its value.
>
> (D, Raval councillor)

What the councillor seemed to ignore is that, as Zukin (1991) explains, the cultural re-evaluation of the city centre is the first step to economic gentrification, which is already reflected in El Raval with its rising rents. The reaction that 'gentrification' triggers amongst planners and politicians shows a reflexive awareness of the negative implications that this concept has in academic and planning literature. However, despite these negative implications in reducing the public sphere and simplifying public life it is regarded as the only solution to restoring the economic and social value of marginal neighbourhoods. And, as El Raval's councillor insists, in order to make El Raval part of the city, a 'social mix' has to be secured. That this is a 'social mix' from which the economically frail are excluded is acknowledged by most official agents interviewed, suggests that a new form of planning is emerging – cynical planning.

Despite politicians' assurances, the frailty of regeneration in terms of its social improvements is questioned by an increasing number of critical voices (Heeren 2002; Balibrea 2001; Delgado 2005). Many of the interviews reflected the planning ideologies which portray the marginal city as a sensuously polluting body that, if not rapidly tackled by cleansing urban planning measures, will infect and destroy its surroundings. Architects and planners viewed themselves as '"doctors of space" . . . who coded and recoded a space to their specifications' as Wilson and Grammenos (2005) poignantly discuss in regard to the regeneration efforts in Humboldt Park, Chicago. The fear of a spreading 'uncontrollable city' explains why the regeneration has been implemented with such a speed:

> If we want this neighbourhood to regenerate we have to open it up to the city . . . It is healthier opening it up, otherwise the protected ecosystem will rot. This ends up being very dangerous and it's a risk to this district. Let's be honest, we can sell things beautifully but we still

> have the highest numbers of unemployed and immigrants from the 'Third World' settle down here because it's their natural habitat. We have to keep up the constant transformation or in three years we can lose what we have achieved in 20. We need to act quickly, otherwise we'll go backwards.
>
> (F, social services)

Despite new designer bars, galleries and trendy clothing shops settling into the neighbourhood, and El Raval becoming part of a cultural consumption zone, its assimilation into the rest of the city through gentrification and regeneration has not been straightforward. The future of El Raval still remains uncertain as most of the social problems remain the same. Unlike Castlefield, El Raval remains densely populated and new meanings and perceptions cannot be easily forced into its socio-spatial structure. The best example of the contested outcomes of this regeneration process can be observed in the activities occurring in the public space surrounding the museum which I discuss in Chapter 8.

Conclusion

Whilst planners and politicians do not make a direct reference to the senses in interviews, I have illustrated that much of the discursive scripting of regeneration is based upon an important sensuous dimension. Sensuous metaphors are drawn upon to legitimate the spatial restructuring of space. Ethical or social issues are subsumed under aesthetic goals. Let me clarify what I mean. In both neighbourhoods aesthetic codes are inscribed into the regenerated public areas, and strategies of social exclusion are disguised through sensuous-aesthetic arguments. The imposition and control of new practices are often disguised under the headings of leisure or culture. The spatial practices implemented by regeneration bodies aim to submit areas to a new definition of public life by manipulating the sensuous geography and thereby attracting or deflecting different social groups. Urban regeneration needs to be understood as deeply ideological because it tries to portray these strategies and practices as inevitable and natural. Indeed, future research needs to establish whether we can speak of a hegemonic aesthetic discourse that is increasingly infiltrating regeneration schemes all over the globe and operates through common spatial strategies.

The dialectic of similarities and contrasts between El Raval and Castlefield lead me to conclude that one could think of the situation as one planner designing two cities. Similar strategies, the 'technology of spatial access' and 'designer heritage aesthetic', are used to produce different place identities – homogenization in El Raval and differentiation in Castlefield – yet are underpinned by a common aim: social and economic restructuring. This implies, as Smith forcefully argues: 'the restructuring of spatial scale, insofar as the fixation of scales crystallizes the contours of social power – who is empowered and who contained, who wins and who loses – into remade physical landscapes' (2002: 435). Aesthetics, culture, liveability and urban regeneration are perversely linked in both neighbourhoods' planning strategies to attract a new public. The consequence is that a parallel social geography is established by the regeneration strategies in both cities. These strategies aim to engineer public life around present and desirable features such as the recuperation of history, design, middle-class practices or cultural consumption; undesirable features such as the marginal, recent past or working-class spatial practices are actively glossed over, hidden or removed from the picture. At the same time planners concede that the danger of regenerating areas is 'always that you sanitize the spaces' and

> that you do something that destroys the character that first attracted your attention to the area and the danger is, in a place like Castlefield, that you change it from an area that is rather gritty, low, as I said at the beginning: the tyre burning, the scrap metal people, the offal processing factory . . . into an area that is more welcoming to visitors. Bringing in visitors can actually destroy the character of an area, so there's a balance to be achieved.
>
> (Planner K, Manchester city council)

In the next two chapters I explore whether this is the case by investigating how regeneration is perceived and experienced by the many different users of these contested spaces in Barcelona and Manchester.

7 Perceptions from 'down below'

> But the city in its corruption refused to submit to the dominion of the cartographers, changing shape at will and without warning.
>
> (Rushdie 1988: 327)

The presence and articulation of official representations tell us little about the ways in which urban transformation processes are perceived, evaluated and lived in the everyday of 'ordinary practitioners of the city [of those that] live "down below"' (Certeau 1984: 93). While the 'concept city', the city of theoretical constructions, can be captured in the form of representations portrayed in planning documents or panoramic postcards, the movements and experiences of those living the 'practised city' create a much less unified picture or easily legible city: 'The networks of these moving, intersecting writings compose a manifold story that has neither author nor spectator, shaped out of fragments of trajectories and alterations of spaces: in relation to representations, it remains daily and indefinitely other' (Certeau 1984: 93). Despite the lived city standing in contrast with the representation of the city, the two are in constant dialectic with each other. Representations, mental conceptions of spaces, organize perceptions. Representations of space 'constitute "stocks of understanding" via discourse . . . They help make "facts" about people and processes as representations of reality lodged into narrative' (Wilson *et al.* 2004: 1174). To put it simply, representations inform people's understandings of cities (Shields 1991, 1996). So, for example, a common trait in many of the interviews held with locals in Castlefield and El Raval at the beginning of this study was a self-reflexive perception of the historical and touristic unworthiness of their neighbourhood. As a Castlefield tourist guide states, before the regeneration there was a general disbelief by Manchester's citizens that Castlefield's or Manchester's built environment could offer anything valuable for outsiders:

> When I used to say in 1980 that I'm a tourist guide, I've had people fall over with laughing! 'Tourist guide'! What do we want one of those for in Manchester? There's nothing here.
>
> (Jane)

A similar perception of El Raval not being historically or culturally 'valuable' was reflected in my experience of tourist walks through the area. During the first few guided tours I took in the neighbourhood in 1998 we were generally greeted, especially in the south of El Raval, with a mixture of disbelief and hostility by residents. These expressions ranged from silent stares and people approaching the guide and asking what it was all about, to youngsters mocking the guide and shouting: 'Yes, we are from the barrio Chino, are you scared? [laughter]', and shortly after an old woman shouting at the group: 'If you want to see nice things go to the cathedral. There you get some pretty sights, there's nothing here!' These experiences demonstrate how wider representations of space filter through and inform individuals' meanings and experiences in particular places.

In this chapter I examine the responses to and perceptions of the regeneration strategies by those who use these spaces daily, such as tourists, new and established residents, shop-owners and others. Users' narratives are understood in terms of Certeau's notion of embodied stories '[e]very story is a travel story – a spatial practice' (Certeau 1984: 115). Particular attention is paid to the ways official representations and regeneration strategies are negotiated in daily life. Experiences of place are not monolithic but constructed in terms of different users' positions and relations to place, as users themselves turn the same space into a variety of places (Pile 1997). The first two sections of this chapter analyse users' competing responses to the strategy of accessibility, and show how their experience sometimes digresses from, and at other times coincides with, official strategies. The last two sections discuss users' reactions to the aesthetic recoding of the neighbourhood.

Castlefield: exclusive public spaces

In regenerated neighbourhoods the already transformed is marked by an absence of contrasting traces of time. From the planners' view regeneration projects gesture the ever present: old historical buildings are cleansed of their physical traces of time and embedded in a contemporary spatial code of 'designer heritage aesthetic'. Regeneration projects are conceived as eliminating the traces of the recent tangible past: the decay, the smells, the dirt that have covered a 'valuable' history of place, and then transforming this decaying landscape into a future image for promotional brochures. To support this strategy former industrial places such as Castlefield are portrayed as a 'forgotten no-man's land': as empty spaces that need to be injected with new life. However, interviews with people who have had businesses in this area over long periods of time resist this representation by offering a different story. Ted, for example, recalls that in his youth during the 1970s, and despite being geographically isolated from the city centre, Castlefield was an active place in terms of its workforce. Liverpool Road had amongst others a barber, a Jewish watchmaker and range of small garages. Castlefield was then 'pretty much industrial, there was nothing targeted to the area. It was business that did not matter where it was. No passing trade' (Ted). An impression of the way Castlefield was before the regeneration could still be gained in 1999 at the corner of Liverpool Road and Deansgate, which contained an area with small car-repair shops, lawyers' offices and industrial workshops. Heavily accented voices of workmen sang out from open garages, their voices backed by the steady clanking on metal and the smell of petrol and brake fluid adding a thickness to the air (Figure 7.1).

Figure 7.1 A small business in the back streets of Liverpool Road. (Photograph by author.)

Most of the people in these businesses had developed social networks over time and workers met regularly in 'Ted and Jerry's', Castlefield's only newsagent, or at Debby's, a 'greasy spoon café', to exchange news and gossip. While most business owners interviewed agreed that changes had to happen to improve the physical decay of the area, most felt that the regeneration of Castlefield had been damaging to them. Rents had been steadily increasing and had driven out many small businesses. During the late 1990s closed businesses and buildings for sale were a prevalent sight in an area where many small workshops had once been. The remaining local businesses resisted what they perceived as the touristification of the area, arguing that Castlefield was 'losing its character' (Kevin). They claimed that the regeneration was mainly catering for the needs of outsiders, constructing a series of entertainment venues such as the Roman fort, the Science and Industry Museum and the Urban Heritage Park. Little had been done to support existing local businesses. In fact, much of the surrounding area had not been upgraded. Kevin, the owner of a coffee shop, argues:

> Castlefield is a place for tourists now . . . You know, the small businesses have been here for years and years and they are part of Castlefield. The tourists are here only for a day. It might be part of the changing times but it is all geared to tourists, which eventually makes something like Castlefield a spectacle for them.
>
> (Kevin)

While one can detect a sense of nostalgia here, it is very different from the discourse expressed in official representations of a romanticized industrial past. During the interview, Kevin expressed a sense of loss for the recent past, for the gradual disintegration of a business community that constituted Castlefield's character during the years of institutional neglect. In Kevin's view businesses showed a more truthful affinity to the place than the one-day visitors who, in his opinion, only perceived the spectacle, the front-stage of Castlefield, while 'we believed in Castlefield, when nobody else did' (Kevin). As discussed in the previous chapter, the strategy of accessibility was to resignify Castlefield as a desirable place for a global audience by developing and stressing its heritage character. The disadvantage of such a strategy is that locals experience such gestures of place as catering only for the needs of a transitory global audience, uprooting Castlefield from its local context and creating a 'spectacle'. The focus on tourism development is interpreted as disregarding local needs, a common feature of regeneration schemes.

This conflict between global marketing and local needs is reflected in the resistance shown by long-term shop-owners and residents to the proposal to rename the area: Castlefield, rather than Knott Mill or Deansgate, the original neighbourhood name. Castlefield's new name was regarded as a 'contrived thing, just a marketing idea' (Ted). It led Ted to describe Castlefield cynically as 'a bit like a star shape, it is sprawling and everyone on the outside tries to say "we are in it"'. In Ted's view, Castlefield is no more than a brand name, an artificial construct that can be expanded symbolically to other areas of the city that benefit economically from this association.

Shirley, an estate agent, observes how Castlefield had been transformed from being a place where 'we had to give people maps and all sorts to get them down there [to look at property]', to 'a well-known area'. Indeed since 2000 approximately 20 new residential developments have been built in and around Castlefield. This indicates the success that access and resignification strategies have had for the property market. Castlefield has become a reputable place to live and have businesses in, especially if they are based in a heritage building. A media worker explains: 'It is quite good when you are introduced to someone and you say where you work. Everyone knows the building in Manchester because everyone knows the area' (John). Needless to say, house prices are unaffordable by those on average wages in Manchester. Supporting Ted's impression, Castlefield's geographical remit is continuously 'stretched a little bit more' (estate agent) by developers. Castlefield, as planned by regeneration strategies, has become a profitable 'status symbol', a brand name associated with expensive inner-city living and known amongst Mancunians as catering for 'a particular segment of society ... It's very much middle class without question' (Jane, tourist guide).

A by-product of the strategic marketing of the area to upper-income groups is that Castlefield's accessibility to a large transient public has not materialized. Castlefield as a public place for a general segment of the public, as an area for leisure and recreation, is only encountered by chance. In fact most people I asked for directions in Manchester's city centre in 2000 could not tell me where Castlefield was, despite it being only a ten-minute walk away. The area is generally not perceived as offering many attractions, except for the museums and the recently closed Granadaland. It is perceived as a difficult area to access because of its lack of public transport – at the time of the study in 2000, the area had just one regular bus and the Metrolink station. To many people Castlefield is considered as 'slightly the "posher side" of town' (Anne, secretary); 'it's like a boost up' (Mary, market visitor); or, as Ted the newsagent explains, 'the rich relation to the city centre'. Many workers interviewed did not know about Castlefield before they started working there, as Daisy, a waitress in the coffee shop, explains:

> The girls that work here don't know it exists, lots of the Mancunians don't even know about the place, don't know anything about it . . . I mean the markets; they only advertise it on the day. So they spend all this time doing it up and they are not really publicizing it for the people living here. I mean in the end it is your city and people should know about it.

Daisy's words hint at one of the key problems with Castlefield as a regenerated public place. While it has been transformed into a more accessible place physically, the commercial strategies and spatial practices aimed at those on higher incomes create a public conception of 'exclusive public spaces'. For many Mancunians, especially those in lower-income groups, Castlefield remains an invisible place in their public imaginary.

In the conception of the city officials the spatial technology of accessibility was aimed at fostering Castlefield's inclusion within the city centre. Complementing this, established residents emphasize the advantages of Castlefield's central geographical position as being within walking distance of the city centre, work, entertainment and shops. Paradoxically, though, Castlefield is not experienced symbolically as part of the city centre's public space, for two reasons. First, it offers a 'characteristically different' environment and atmosphere to the city centre. Residents evoke in their descriptions of Castlefield suburban gestures of place which are often mentioned positively as their main reason for settling in this area. Emphasized in these narratives is the quietness and perceived safety of its public spaces as well as its 'garden aspect' that makes it feel like 'an oasis in the city' (Sarah). Locals and tourists alike reflect on the contrasting spatial practices afforded through the design of the area which consist of leisurely walking in a mainly empty area and contemplating the Urban Heritage Park, when sitting in one of the European-themed café bars: 'the centre is completely different, because it is so busy and full of shops. It is very different here. It is very quiet here' (tourist from London). Castlefield's silence, lack of activity and broad open spaces produce a sensuous difference to the city centre and create its distinctiveness. Yet its detachment from the city centre and the lack of commercial facilities leads to the second, more negative, assessment of Castlefield's symbolic difference to the city centre: 'there's no bus service into the city centre; there's no telephone kiosk; there's no post box. There's none of the facilities of the city centre. You are not in the city centre at all. You are deluding yourself if you think you are' (Mike, established resident). Interviewed residents all agreed that Castlefield lacks a sense of being a residential area because its public space does not provide the expected service infrastructure of a city. It also does not provide any facilities for families to settle such as schools or doctors: 'there's no schooling here, you see couples with children have to leave by then . . . This place is not designed for children' (Vicky, established resident). One resident concludes: 'You sometimes feel like a tourist. Everybody else is a tourist. It has no sense of living really' (Kate, established resident). Indeed, it was only in 2000 that the first upmarket delicatessen opened in Castlefield (Figure 7.2).

Long-term residents feel that Castlefield is increasingly catering for a transient population. When first moving into new developments in the mid-1990s most residents felt a strong sense of community, triggered by a feeling of being 'urban entrepreneurs'; 'it was innovative, you felt you were doing something different to live in the city' (Sarah). This feeling has levelled off

Figure 7.2 Castlefield's first 'corner shop'. (Photograph by author.)

as the novelty of city living has decreased. Most residents are only acquainted with their immediate neighbours and generally regret the quick turnover of population which does not foster close relationships. While not manifestly resisting the transient character that has been promoted by the regeneration strategies – in fact some residents choose to live in Castlefield for its 'anonymity' – these experiences tell us something about the 'sense of place', and the social relations that are promoted in Castlefield as a public space. An exclusive public space for outsiders, for insiders, it is a public space reminiscent of that within gated communities. Reflecting Rifkin's (2000) description of North American gated communities, it is an area in which property owners buy their rights into the lifestyle of a preselected homogeneous group. The bonds established are focused on one's own property rather than on a broader geographical area or 'community', as this resident explains: 'From my experience people in Castle Quay cared about their particular building but not really about the rest of the area as long as things looked OK and were clean' (Janet, established resident). From the analysis one can conclude that Castlefield has the potential to become an 'exclusionary enclave' (Marcuse and van Kempen 2000), an insulated area of upper-income residences.

El Raval: invisible public spaces

Improving the economy of access and changing El Raval's negative reputation in the public imaginary was a main concern for its residents. However, this was based on premises different from those described by the local officials. Residents' main expectations of the regeneration have been that this process will 'get rid of this blight' (Oscar) that outsiders associate not only with the neighbourhood but also residents' identity. Oscar, a 30-year-old resident who was born in El Raval, reflects on how outsiders conflate the physical decay with the people living in it, regarding them as 'Third World people' separate from the rest of Barcelona:

> A few years ago when you said you were from El Raval people would look at you in a funny way. That still happens when you're of a certain age. Living here as a young person is one thing, but if you're a 40 to 50 year old person people look at you funny. Therefore we have this kind of complex, as if we were Third World people, as if is the shabby neighbourhood of Barcelona, that has to be got rid of.
>
> (Oscar, established resident)

Oscar's comments show how many of the negative territorial discourses and representations promoted by the regeneration strategies have permeated residents' self-identity in El Raval. As Oscar reflects, regeneration discourses, with their aim of attracting young people to the area, have produced different social spatializations in El Raval. Young people are identified with the invigorating forces of regeneration, while old and established residents are conceived as still belonging to the old marginal value systems that the regeneration is so eager to clean up.

Official representations construct criminality and marginality as key defining features in El Raval's public spaces, and therefore aim at a dilution of its public life by cleansing the area physically and socially. Residents resist this uniform representation through two central discursive constructions. The first discourse portrays 'el Chino', the 'vice-ridden area' as an urban myth, a *mobile space*, which is everywhere and nowhere. Hence, when interviewing established residents, most of the time it is difficult for them to geographically locate 'el Chino'. For residents, the neighbourhood itself is divided into smaller quarters, and 'el Chino' is always located outside their immediate familiar area. This constant relocation of the marginal can be understood, as Sibley argues, through 'a desire of those who feel threatened to distance themselves from defiled people and defiled places' (Sibley 1995: 49).

The second discourse consists of romanticizing the criminal past. While residents do not dispute the fact that marginal activities did take place in the neighbourhood, until the introduction of heroin[1] these are portrayed as taking place peacefully, in parallel with the everyday life of the neighbourhood. In these discourses heroin is constructed as the scapegoat that destroyed a carefully orchestrated sociability and is made responsible for the extreme social and economic decline of the neighbourhood. Established residents make a clear distinction between the 'golden past' before the entrance of heroin, which they characterize as a time of social cohesion, despite – and probably because of – the precarious living conditions in the area, and the time after the entrance of heroin, when the community was destroyed as relationships broke up and the neighbourhood became a seemingly dangerous and lawless place to live in. Maite sums up these feelings in the following way: 'Until the '80s we were "el Chino" but with pride; after the '80s we were only the Chino, and that is when we started saying we don't want el Chino we don't want this' (Maite, established resident). Hence, 'accessibility' by residents is conceived of primarily as changing the neighbourhood's negative reputation by specifically eliminating the drug-related criminal element from the area so that outsiders will start to enter the neighbourhood. Within this context El Raval's new cultural venues are regarded as crucial in reframing the reputation of El Raval within Barcelona and attracting its citizens into this once no-go area.

Changing the reputation of El Raval is taking longer than expected, as Julia and Majid point out. When interviewed in 1998 they had been trying to sell their newly built flat for over a year. While lots of people came to view it, 'they would say that they like the flat but that it is in El Raval. People have lots of prejudices, they aren't convinced it has changed and it is difficult to convince them' (Julia, new resident). A common argument interiorized by many residents is that it is El Raval's run-down, oppressive sensescapes, still dominant in many non-regenerated areas, that place it outside mainstream society. Hence, Gema, born in the neighbourhood and owner of a restaurant, believes that the reputation of El Raval will only change when the physical and social landscape has been completely remodelled. She argues: 'Nobody wants this neighbourhood when ethnicity and marginality is so little balanced. Nobody wants to be near the marginality because marginality and poverty is ugly. Nobody wants to live next to ugliness' (Gema). These perceptions support the official representations of the non-regenerated as 'uncivilized' and use space as a hierarchy to distinguish 'civilized newcomers' or gentrifiers, from 'uncivilized newcomers' or immigrants. Gema locates El Raval's new racialized minorities in the social category of the 'marginal' and correlates them and the places they inhabit with negative sensescapes. As Sibley explains: 'Portrayals of minorities as defiling and threatening have for long been used to order society internally and to demarcate the boundaries of society, beyond which lie those who do not belong' (1995: 49). Yet, immigration in El Raval is a relatively recent phenomenon, having only started in the mid-1990s. El Raval has been shunned by mainstream society for much longer. Similar to heroin at an earlier time, it is now immigrants that have become the scapegoats for El Raval's reputation.

Complementing the officials' discourse for a 'purification of space' a common argument amongst established residents is that access into the area will only be guaranteed by a new environment that displaces El Raval's gridlock of narrow, poorly lit cobbled streets with broader, ordered, paved and well-lit streets. One of the first demands made by an independent business association,[2] and implemented by the city council, was to create a circuit of 'pleasant' pedestrian streets through the northern part of El Raval that are expected to dispel outsiders' representations of danger and fear. This was an idea that 'was based on our neighbourhood being an unknown neighbourhood because people were scared to walk across its streets, because they did not know it and saw it with a bad reputation, although that isn't the reality' (Miriam, independent business association). Transforming the affordances of places to convey a sense of safety is a common practice in regeneration strategies (Newman 1972; Davis 1998). It is one that reflects Bauman's (1999) argument that a way of dealing with fear in late modernity is to focus on physical features that can be changed to make a place appear 'safe', rather than tackling social structural problems which would take much longer to change.

While everyone agrees that pedestrianization makes the area more easily approachable for tourists and outsiders, local residents question whose benefit this has been done for: 'Now it sometimes seems that they are making a neighbourhood to visit but not to live in' (Oscar, established resident). When asked to comment further he states:

> If you're a normal person who works and has to leave by car in the morning to work, it is becoming more and more difficult. Not only finding a parking space – that happens in the whole city – but also in this neighbourhood real brutal things are happening. Now almost all streets are pedestrianized. That's all very well, but that's a neighbourhood to go for walks in not to live in. Do they think that all the people who live here are foreigners[3] and go on bicycles? *Hombre!* Most people here are 'normal' people not students, foreigners or old people.
>
> (Oscar, established resident)

Oscar exemplifies here a conflict of interests played out in public space. On the one hand, there is a clear policy of touristification of El Raval which involves providing a landscaped environment that caters for a transient consumption of place. On the other hand, there are the locals' everyday functional needs.

El Raval's new access routes, and the resignification of the area as a cultural quarter containing museums, libraries and research centres, are regarded positively by both new and old residents as they attracts outsiders. Residents generally view increased visitor numbers in El Raval's cultural quarter as a reflection of its changed reputation and link this with a re-designed urban landscape. As Majid, a new resident, states, the public's change of perception is reflected in people saying:

> let's go to 'El Raval' and not 'el Chino'. It has been oxygenated. You can breathe now. It was a very dense neighbourhood and now it is less so. Now it is open. You can actually cross it. Before people said: 'we won't go through El Raval after 8 or 10 o'clock because of the prostitutes, the crime, and the drugs'.

Yet despite more people venturing into El Raval, neither residents nor shop-owners regard El Raval symbolically as part of the city centre. In their perception it has not managed to attract Barcelona's international tourism, nor is it considered as part of the city centre walking and shopping circuit. Nuria, vice-president of the official residents' association, comments that outsiders who visit the new city facilities do not seem to realize that they are in El Raval, a fact that she resents. Voicing a common feeling amongst established residents, Nuria resists the homogenization of El Raval's public spaces with the rest of Barcelona promoted by the council and expresses a strong sense of territorial ownership:

> People don't think that 'el Chino' is the Central Library of Catalunya, is the Liceo [opera house], is the MACBA, is Sant Pau del Camp [a Benedictine foundation of the tenth century]! Nevertheless they come to marry here, they come to study here and if you ask them: 'Did you go to the Central Library?' 'Yes.' 'And to which neighbourhood did you go?' 'To the centre.' 'No, you did not go to the centre, you are in MY neighbourhood, in El Raval.'
>
> (Nuria, official residents' association)

My own ethnographic observations show that visitors tend to access El Raval through regeneration corridors, rarely venturing into the non-regenerated backstreets. Here one encounters the chink of wineglasses and excited chatter pouring from stylized bars and trendy restaurants. Inside a young crowd sips Martinis and other cocktails. Next to them are exclusive boutiques and art galleries, their windows displaying selective items from

Barcelona's up and coming designers or artists. Both new and established residents condemn how visitors to these venues seem to be oblivious of El Raval's everyday reality. These places are lifted out from their immediate local context to become part of a global network of trendsetting urban cultural capital connected through lifestyle and promoted by travel and design magazines. The public space outside these cultural circuits becomes 'invisible' for particular groups of society. While in Castlefield we have *exclusive public places* that do not attract lower-income groups, in El Raval we have *invisible public places* that do not attract higher-income groups. The reverse is also true: hardly any established Catalan-Spanish residents I interviewed had ever visited any of the new restaurants, shops, or cultural venues in their neighbourhood. We are witnessing here a segmented public space in which different groups make parallel uses of a place but do this without engaging with each other.

While official discourses emphasized the function of El Raval's regeneration to physically connect its public spaces to the city's centre, new and old residents alike already experienced El Raval as a central location. El Raval is well supplied with a variety of grocery shops, markets, bars and restaurants that make it a self-contained and 'easy neighbourhood' (Judith, new resident). In fact many of the old people interviewed had not left it in years. On the other hand, it is within a few minutes' walking distance of Barcelona's key symbolic centres such as the Plaça Catalunya, Las Ramblas, the main shopping areas and the sea. All these factors make El Raval an attractive area for private developers, and have famously led the president of the official residents' association to say: 'El Raval is land of gold built with mud houses', referring to the disparity between the desirability of El Raval's land for property and the poor quality of its housing stock. Old established residents are at times puzzled by the array of contradictory messages they have been getting since the start of the regeneration. On the one hand there are negative official representations and press coverage of El Raval, on the other private developers are hunting for their properties:

> Nowadays you get home and you find leaflets that say that they will buy your flat for cash. And that's what I don't understand, we have to leave this neighbourhood because they tell us it is not a good one, but on the other hand new people are coming in, I don't understand that. Despite what they say we couldn't have it better here: we've got everything here. You don't need to get the underground, the bus or car.
>
> (Maria, established resident)

As Barcelona has entered the global property market in recent years, real estate 'mobbing', often also referred to as 'real estate and urban violence', has increased in areas such as El Raval. Vulnerable social groups have been pressured into leaving their properties through a range of legal and illegal strategies, ranging from intimidation to extreme rent increases[4] (Heeren

Figure 7.3 'No flats for sale – OK?' (Photograph by author.)

Figure 7.4 'Flat for sale'. (Photograph by author.)

2002; Union Temporal d'Escribes 2004). An increasing number of residents' groups are voicing complaints about the speculative processes that are reshaping the city of Barcelona and are signalling the end of a passive consensus on the development of the city (for example, *La Veu del Carrer*[5]). In fact some residents are so tired of the constant harassment by estate agents that they put notes on their doors stating: 'No flats for sale' while others are eager to take advantage of the speculative process in El Raval and advertise their properties.

Castlefield: promoting selective histories

In a regeneration process the co-presence of different times and spaces is emphasized by the spatial transformations and tensions that emerge between the transformed, the old and the envisaged. The most pronounced physical changes in both areas are constructed discursively as a move from a darkness which is linked to dirt and disorder to an experience of light[6] which is linked to cleanliness and order (Stallybrass and White 1986). However, due to the different socio-cultural settings of each neighbourhood, the recoding through a 'designer heritage aesthetic' in El Raval and Castlefield produces different responses and themes within each area.

A common theme in the interviews held with the different user groups is a reflexive awareness that Castlefield's Heritage Park offers a range of *selective histories* ingrained in the sensescapes of the place. This is reflected in the commercialization of history in which the transformed vernacular architecture has become an important selling point for developers. As an estate agent, explains: 'when we sold Slate Wharf they actually put a bit of history in the brochure' (Figure 7.5). History is the surplus value that one gets when buying a flat or office in the area. In these discourses, history is relegated to a decorative role, as a visual backdrop and one which promotes the consumption of landscape: 'I don't hear of people that say "Well, I want to go and see the Roman wall", or "I want to go and see the history in Castlefield", but they like what has been left by the history there' (Shirley, estate agent).

Yet whose history is told in Castlefield? Official discourses equate the physical decay of Castlefield with a 'wasteland', 'devoid of positive social, material and aesthetic qualities . . . purely an abstracted and quantitative entity technically identified by the assumed absence of activity or function' (Edensor 2005: 9). Most established residents and shopkeepers in the area, however, challenge the normative aesthetic coding of the regeneration and do not link the industrial decay of the area with negative associations. Their memories, while highlighting the desertedness of the place, evoke the human history ingrained in the affordances of the past sensecapes. This

Figure 7.5 A sales brochure advertising Castlefield's heritage.

Source: Deansgate Quay Development, Crosby Homes 1999.

contrasts with the planners' and politicians' aims of erasing any trace of the industrial decline. The dense sensuous industrial landscape, represented negatively by officials, is subverted in descriptions as 'honest rubbish'. The gestures afforded by the dereliction, the dirt and dust, provide a sensuous proof, a credible account of the passing of time, and account for a lived past inscribed in the place, rather than portraying desolation. Lacking officially prescribed uses, ruins permit a large range of practices as Ted, who grew up in Castlefield, explains:

> I used to wander around here, all these places were derelict. I remember going to rooms covered in dust, the tables set up, cups on them as if people had left in a hurry. Tramps were living there . . . I had a garden [laughs], it was all this [makes an embracing movement with arms]. I don't know, it was great, I could always find something to throw in the water or to climb on. Things were a lot grimier, a lot of the brickwork is sandstone and has been sandblasted, it was big piles of honest rubbish, brickwork, heavy masonry.

Other long-term residents and people who have seen the area change over time also resist official discourses by attaching positive memories to the sensuous decay of the industrial city. The pungent smell of burning rubber tyres, the overgrown cobblestones, the cyclical humming of a cement mixer are evaluated as legitimate senses reflecting the working patterns of the area. These sensescapes are contrasted with the atmosphere that Castlefield acquired after the regeneration, regarded as bland and sanitized, marked by an absence of sensuous contextual markers:

> It doesn't have the smell it used to have. The smells related to the jobs that were going on there. The cutters on the scrap metal. Years and years ago, whatever industry was there, and whatever was handled on the water. And you had all the dirt from the coal barges . . . [Now it smells] very neutral I think. Unless you get somewhere near the café bars, when they've got a lot of coffee going. There's not the atmospheric feeling of Castlefield, it's very bland.
>
> (Jane, tourist guide)

Jane seems to express here a sensuous feeling of loss. One can draw parallels with Benjamin's (1975) argument that aura is disappearing in the age of mechanical reproduction. Indeed most interviews express concerns about the regeneration's 'finished product' (Vanessa, established resident). In this case, we can see how residents perceive regeneration as a force which engulfs an area and leads to a commercialization of place, where buildings do not stand for what they were but are landscaped to become scenes of an exhibition or theme park. Signs explaining the history of these places embed the buildings into a normative historical narrative that celebrates some aspects of history but not others: 'Now they've put up flowers, nice lamp-posts, and nice signs. I mean they have aestheticized the place, they have made it look much prettier. And it is very noticeable . . . you got those nice blue signs that say Castlefield and all the posters' (Kate). Some established residents recognize this representation of history as a partial, selective memory of particular activities and histories of the place:

> you do see this quite remarkable nineteenth-century industrial architecture. I'm not talking just about the buildings, but the bridges, the railways . . . Yet, this was also a horrible place, I mean just 20 or 30 years ago and probably even worse 50 years ago. And that is an aspect of Castlefield that has been sort of wiped out from historical memory. If you look at all the plaques and that, it's all about some time in a romanticized version of an early nineteenth century.
>
> (Jim, established resident)

Yet, while questioning the artificial manufacturing of Castlefield's history, residents simultaneously experience the 'look' of the place as one of its crucial identity markers and describe it as the main reason for living in Castlefield. Complementing official representations some residents cherish most the visual staging of history, the 'visual imperialism' at play:

> I like the urban architecture. I suppose the noise of the trains is a bit irritating but I just like it, particularly at night. It's attractive seeing the trains and trams go by and it's all lit up and there's all this Victorian brickwork and the canals and reflections in the canals. It's the look of the place I like.
>
> (Vanessa, established resident)

However, the various historical themes in the area are judged differently. The Roman fort, for example, is perceived as a fake, because it was not an original leftover of the physical landscape but is a replica and therefore it

> just looks false . . . it looks modern, it looks far too modern. Well, we know the site was here but it's too modern. It's somebody in some time or other that has built a mock Roman fort. I'm told they didn't use sandstone but used wood, let's say for the sake of the argument.
>
> (Mike, established resident).

The Roman fort is regarded as a marketing strategy to attract visitors; it is part of the tourist circuit in the area that takes people around the area's urban heritage (Figure 7.6).

Castlefield's pre-regeneration landscape is remembered by many residents nostalgically. In their view the regeneration has led to an increasingly sensory-neutralized, circumscribed environment in which the visual sense

Figure 7.6 The Roman fort replica. (Photograph by author.)

dominates the others: 'It was a working canal then, now it is a little bit more touristy' (Rob, established resident). The past gestures of place where the different senses worked in a sequential order, so that each sense offered an equal amount of information, are longed for. The ruins, the wildness of the place were perceived as mysterious, filled with multiple meanings that allowed for the unexpected to happen and providing an alternative aesthetic. These are 'spaces in which the interpretation and practice of the city becomes liberated from the everyday constraints which determine what should be done and where' (Edensor 2005: 4). These experiences are evoked in an almost sentimental manner:

> We used to go before it was all done up and it was really rough and no lights down the tunnel. We loved to walk down there from Castlefield up to the Canal and Piccadilly and there was ice on the canal. And we went down Deansgate with torches; it's quite a long tunnel, no lights in. We suddenly heard a scattering and we went: 'What the hell is that?' We thought of huge rats or whatever coming down the tunnel but it was just someone throwing a brick on the ice from the bridge, it just skidded along the ice. I think it had more character before, this is really unfashionable of me to say. I like old bricks, before they did it up, put light in it, sanitized the floor; it used to be just a mud floor.
>
> (Rob, established resident)

Ruins are read for the dialectic of absence and presence of human life that they project (Simmel 1959). They are experienced here as offering a space for imagination and fantasy. This feature is taken out of the regenerated landscapes which provides only one normative meaning and thereby reduces the multiple readings of the city. Non-regenerated places are abandoned places, yet the past's presence is evoked in the sensuous markers of time, in the traces of human use that, for example, an indentation in the floor or dust-covered cups might suggest. As Simmel explains:

> In the case of the ruin, the fact that life with its wealth and its changes once dwelled here constitutes an immediately perceived presence. The ruin creates the present form of a past life, not according to the contents or remnants of that life but according to the past as such.
>
> (Simmel 1959: 265)

It is from this premise that we need to understand the opposition of established residents to a 'complete regeneration' of the few remaining old shops, such as the newsagent on Liverpool Road. 'Ted's and Jerry's' has been in Castlefield for over 30 years. In fact, Ted's father opened the shop in the early 1970s. The cluttered inside has not been renovated for years and its aging shelves strain under an overwhelming array of drinks, sweets, cans and stationery.

Figure 7.7 Ted and Jerry's newsagent. (Photograph by author.)

The official aim of telling a fabricated selective history is contested and questioned in some interviews, as one can appreciate from Sarah's words:

> for instance 'Ted's and Jerry's' as we refer to it. Those guys are ancient and their shop isn't, you know, high standard. It works, if you run out of milk you can go there and they're open reasonable hours. But it's this sense of: should everything be tidied up and made pristine? Because . . . those would have been back-to-back slums at one time. And there was a huge epidemic of typhoid in that area, where something like 20 odd thousand died, because of the terrible conditions and everything. No trace of that is left now. I'm not saying that we need to keep it, but that's even erased from people's memories, people coming in as visitors wouldn't pick up on any of that information.
>
> (Sarah, established resident)

Likewise, in El Raval, newcomers remark that the difference between new public spaces and non-regenerated spaces is that in the latter the social history can be felt as ingrained in the walls and therefore 'it is an authentic neighbourhood, you can see its history everywhere' (Albert, new resident). However, as a French tourist remarks, the newly regenerated places could be 'anywhere'. Thus:

> [w]hen you see the MACBA, you don't see history, you see an architect that has made a building and that is it. But in the rest, you can see the work in the stones, in the houses full of history and you can see how an artist has worked at that.
>
> (tourist)

Hence one can conclude that the official's 'designer heritage aesthetic' is subverted, as it is perceived as purposefully manufactured and missing 'past authenticity', i.e. the human traces of lived history. In Castlefield and El Raval the past becomes, in Williams' (1965) sense, a 'structure of feeling' in which places are remembered by the experience that they evoke, 'revealing how a city – like an epoch – might have certain predispositions of sense and sentiment' (Pile 2005: 3).

The tourists interviewed in Castlefield focus their place experience very much around the Urban Heritage Area and the information provided by the signs that guide them through the area. In fact all the tourists interviewed describe the enjoyment of walking around a historically significant place. As 'post-tourists' (Urry 1990) they self-reflexively describe the Urban Heritage Park as a 'stage' that helps them to imagine what the place had been like before. In fact, the tourists interviewed point out that the silence and emptiness of the area, the sensuous difference from the city centre which is described as busy and loud, is what makes it easy for them to imagine what life had been like before. For the self-reflexive post-tourist silence becomes the carrier of presence. A German tourist explains:

> Because it is so quiet it makes it easy to imagine the things that were going on here in the nineteenth century and the information of the centre is very good and complete. It's so different from modern life, it really gives you a good impression of life in the last century.

The interviews suggest that different user groups appropriate the same space in a variety of ways. Users show a reflexive awareness of the manufactured character of Castlefield's history in that they regard it as representing *selective histories*. These findings point out that, as Urry (1995) argues, consumption is an active process, in which consumers do not passively accept hegemonic meanings or uses but adapt, transform and reshape them. The ways that various user groups 'consume' the area create an array of meanings of place which construct multiple associations around the different sensescapes afforded in Castlefield.

El Raval: disrupting social lives

The contrast of different times is much more perceptible for the walker in El Raval than in Castlefield. As it is a densely populated area the process of regeneration cannot occur uniformly, but must make incursions in different points and develop at different speeds. Present, past and future are dramatically contrasted in the sensecapes of the area. Smooth cream-coloured designer buildings stand alongside elaborately designed nineteenth-century grey facades. On the opposite side of the road, amidst the rubble of half-demolished buildings, stand huge advertising hoardings portraying the future image of the neighbourhood. Residents often refer to their neighbourhood as a place under siege and it is easy to appreciate why. A walk through the derelict streets offers a vista of gutted houses and lonely walls, immediately reminiscent of a war zone (Figure 7.8).

Pink flowered wallpaper rustles in the wind, the shadow of a bed frame marks the tiny bedroom, and the blue tiles of a washbasin are all that remain of the communal toilets. But if this is a war zone, then it is an ongoing conflict, as the ravaged landscape is constantly accompanied by the intense sounds of construction, foretelling its own story of new apartment blocks and broad avenues. Through it all, adjoining neighbours stand silently on their terraces to observe the demolition. And this is perhaps the most emotive scene, as in the process of dismembering El Raval's physical past, we witness the destruction of a living social history, something altogether different from the conserved history of churches and charity houses that will remain. The romantic gesture that the narrow streets might have evoked to the passing visitor is destroyed when faced with the living conditions inside these houses: 'That the outside walls are gone is as if the houses had opened

Figure 7.8 Leftover walls. (Photograph by author.)

their shells and you can see the soul of the neighbourhood, you can see how life really was inside all this' (Carmela, artist).

By demolishing the houses, suddenly the memory of the people who lived in these places becomes alive, albeit fleetingly, soon to be buried under new construction. Hayden's (1996) argument about the 'power of place' springs to mind when she argues for the power of ordinary landscape to nurture citizens' public memory, the power of place to encompass shared time in the form of shared territory. Here this 'power of place' is gradually replaced by a new geography, new gestures of place, that disrupt the shared time but might open up to new forms of shared territory. The official residents' associations, in tune with the discourses of the city council, regard the physical changes as beneficial and as improving the social life of the neighbourhood. However, rather than regarding the improvement of the physical environment as 'civilizing', they perceive the transformations as dignifying one's surroundings, and thereby enhancing the dignity of El Raval's residents:

> I think that the regeneration has given the neighbourhood a breathing lung. The fact that they've demolished housing, that they've resettled families in restored housing has not only changed the physiognomy of the neighbourhood but even the physiognomy of people ... That is because personal dignity starts by dignifying the social environment.
>
> (Nuria, official residents' association)

Residents themselves describe a new feeling of spaciousness that is undoubtedly linked to the demolition of housing, the creation of open spaces and the opening-up of streets, which is linked to a perceived decrease of marginality. Established residents cherish the fact that new affordances in the built environment facilitate better basic living conditions, as this quote by a 70-year-old resident illustrates:

> I like the changes because now we can see the sun, the moon, before we could not see anything ... I have lived here for 20 years and I could never enjoy the sun like I'm doing now, never, I was always stuck in freezing conditions, for sun I had to go to the beach.
>
> (Paco, established resident)

Most local residents emphasize the institutional neglect of El Raval. This is often associated with the physical and sensory downgrading of the place during the pre-democratic Franco years and the perception that the neighbourhood has always suffered from being regarded as unworthy by the city's bourgeoisie. As explained earlier, for over one hundred years, since the new city expanded outside the Old City in 1759, El Raval had been continuously threatened by demolition: 'The regeneration is not a recent process. I know people that have been 70 years without paying rent. What I mean is that we've always had the sword of Damocles hanging over us, waiting for the regeneration' (Nuria, council-run residents' association). Having 'the sword of Damocles' hanging over whole areas meant that house-owners were legally expropriated and therefore discouraged from investing in the maintenance of buildings. People in El Raval have lived with a chronic uncertainty, sensuously ingrained in El Raval in the physical neglect of the place, as Rafael, the owner of a stationery shop reflects:

> There wasn't anything here. In terms of cultural or citizens' attractions, there wasn't anything here. You had the Charity House but it was a big, rambling house, so antique, so old, bad smelling, it smelled old, like dirt, of abandonment, it was for a long time abandoned.

Thus anger and resistance emerged amongst established residents against

Figure 7.9 An old man observing the demolition of the neighbourhood. (Photograph by author.)

the rapid and radical regeneration of the 1990s which bulldozed over a third of the neighbourhood, deemed as 'unworthy':

> The regeneration could have been done without having to bulldoze it all. The bulldozer is necessary when the houses could not stand up anymore, when there's no other way. If at the time the property owners had got the help and courage to invest in their housing we could now be a well-off neighbourhood with small businesses and so on. But we haven't got anything here.
>
> (Gema, established resident)

The above quotes provide us with a different picture of El Raval's 'malaise' than that presented by politicians or planners. Rather than blaming 'the symptoms [as] the cause' to rephrase Smith's (1999: 188) words, what these residents' perceptions illustrate is the lived experience of an area that has suffered decades of chronic disinvestment. El Raval's marginality was not self-inflicted by its residents, as portrayed in the official discourses, but is experienced as a conscious neglect by urban politics.

Overall, users consider the transformation of the physical environment to be generally positive since El Raval had reached an extreme point of deprivation and marginality, and the only way to improve the neighbourhood was with quick and radical measures. Therefore changes are accepted as inevitable, even though the changes occurring are often not regarded as fitting the character of the neighbourhood, clashing with the gestures of place: 'considering the housing around the MACBA, the MACBA seems like an island' (Albert, new resident). The strong sensuous contrast between the new elements such as the luminous Museum of Contemporary Art and the back balconies of surrounding buildings filled with household goods and drying washing (Figure 7.10) leads a local baker to conclude:

> It has little relationship to the neighbourhood. Those of us who have lived in this neighbourhood all our lives and know what it looked like before – you can see that in the back areas of the MACBA – well; there is no relation. But it is good because they are breaking a bit with the atmosphere that the neighbourhood had before. It's better to give it a character of study, of culture than giving it the character of prostitution.
>
> (Javier, baker)

The above example illustrates how the residents' representations both subvert and complement the official strategies. While Javier agrees that the cultural quarter has little in common physically and symbolically with El Raval, the changes are accepted, as the introduction of these new public facilities are expected to resignify the neighbourhood and finally erase El Raval's marginal reputation. Power often works by silencing different possibilities, reducing options. Residents of El Raval were never offered a choice between different possibilities for the regeneration of the area; hence residents regard the cultural regeneration as positive and the only solution, 'because they can see or imagine no alternative to it' (Lukes 1974: 24).

While Castlefield's residents may disapprove of the aesthetic physical transformations, in El Raval people are far more deeply affected by the regeneration which has destroyed their social networks. As all members of residents' associations interviewed comment, despite an effort to ensure that residents were reallocated within 100 metres of their old dwelling, the rehousing of people into new homes has disrupted the sociability patterns of the El Raval. Neighbours were not rehoused together to make sure 'that at least a resident does not lose her social and familiar surroundings, because for many older people, neighbours are not just neighbours, they are like family members' (Nuria, official residents' association). Instead long-time

Figure 7.10 Old buildings surrounding the Museum of Contemporary Art in El Raval. (Photograph by author.)

neighbours were dispersed and rehoused separately, breaking up social bonds and relationships that had been shaped over generations. In El Raval neighbours were in many cases extensions of the family. In this way, neighbours participate in family events from birth to death over a time period of two or three generations in some cases. Moreover, the structure of the housing: thin walls, communal staircases, communal drying areas for washing at the back of houses and dense living conditions overall led to neighbours being acutely aware what was being said, done, cooked and listened to in neighbouring flats, blurring private spaces. In other words, the material geography was intricately folded in with the social geography in El Raval. As Nuria explains, in the view of the residents' associations,

> the regeneration has neglected a fundamental theme: it hasn't been made from a human point of view . . . So I've seen a lot of pain, pain when they had to leave their houses, I've seen people cry, and that hurts because they don't only demolish their housing but their histories.
>
> (Nuria, official residents' association)

The new buildings and streets are celebrated for their sanitation and for offering a better quality of life, but they are not accepted as fitting aesthetically and socially into the neighbourhood. Miriam, a member of an independent residents' association, describes her experience of the new street Aurelia Campmany: 'I was there the other day and I thought I wasn't in my barrio. I looked from one side to the other and it wasn't my neighbourhood.'

Miriam's words draw attention to the importance of the sensuous geography of place informing our experience of 'dwelling in public'. The sensory landscape creates embodied emotional ties and points of attachment between ourselves and the physical environment. Miriam's reaction illustrates the power of regeneration in reshaping the afforded sensescapes of an area which disrupt people's 'sense of place'.

Indeed, in the late 1990s a debate erupted in the local media and amongst intellectuals and architects such as Oriol Bohigas about the poor building material and design of much of the new public housing in El Raval. This debate left residents feeling as though they were not valued, and were being relegated to the bottom of Barcelona's urban pecking order. Graffiti that appeared in the neighbourhood during that time stated 'We want decent housing' and 'No to the cemetery neighbourhood', and further voiced dissatisfaction about the way the regeneration process was tackled. A 75-year-old resident sums up the views of many residents interviewed by stating: 'The new houses resemble graves.[7] I get the feeling that they make them for

Figure 7.11 'No to the cemetery neighbourhood' graffiti. (Photograph by author.)

the old now and tomorrow they will bulldoze them and build new housing there' (Pepa, established resident).

Critical voices from independent residents' groups have been cynical about the physical changes and resist the official representations that the physical transformation of the place leads to social improvements. They believe that the social problems have either been covered up or pushed out by creating designer environments; the real needs of the place have been pushed back-stage. Implicit in the next statement is the criticism that the architect did not know the sensescapes of the place and could not read the gestures of the place that inform the character of the neighbourhood:

> I believe that too often the views of the hippest designers and architects predominate but the reality is a different one. When they set the floor for this square for example, the architect wanted the square to be this way, well he should have known the neighbourhood before taking that decision. I like design but I sometimes get very angry because with so much design they sometimes get it wrong and they do things that are wrong, that are not practical and the life of the barrio needs other things, not so much design.
>
> (Miriam, independent residents' association)

A telling example of the above argument is the anger at 'this paranoia on the part of architects to get rid of balconies' (Oscar, established resident). The lack of balconies in new buildings is deeply felt as it is regarded as a typical feature in the neighbourhood that shapes the relationship both new and old residents have to the immediate urban environment. The balcony helps to 'open up' the neighbourhood, as it permits one to step outside and get a sense of space in this densely built place. A local journalist reflects on the implications of these physical-aesthetic changes:

> I was walking yesterday along the Calle Augustin Viejo and they are demolishing old buildings and constructing new ones. The contrast is brutal. There's the old building with the gas canister on the balcony, an old lady leaning out of the window. It's an old building but it is open because the windows are open, the balcony can be seen, the entrance door is ajar, it is transparent. Next to it you have a stone building, glass and metal, perfectly rectangular, geometric, with automatic doorbells, the windows hermetically sealed, without balconies. They are more interested in aesthetics than utility.
>
> (Ruben, local journalist)

The above quote highlights important issues. First, the new buildings promote boundaries rather than transitional spaces that encourage multi-sensory encounters. This physical change impacts on the social life of the streets and the sociability patterns of residents. An important 'contact point'

Figure 7.12 Old balconies in El Raval. (Photograph by author.)

(Sennett 1996b) with which people in El Raval can enter social relations with each other disappears. Private and public space are now clearly separated; chance encounters are annihilated. As Sennett points out: 'By controlling the frame of what is available for social interaction, the subsequent path of social action is tamed. Social history is replaced by the passive "product" of social planning' (1996b: 96). The second issue is that architects are imposing their vision of bourgeois privacy or 'ideology of intimacy' and

Figure 7.13 New balconies in El Raval. (Photograph by author.)

creating buildings and places 'that do not suggest in their form the complexities of how people might live' (Sennett 1990: xi). Ironically, the inhabitants of these new buildings subvert their new environments by hanging out washing and personalizing their windows – despite the lack of space.

In El Raval, a densely inhabited neighbourhood, the social consequences of the physical transformation have been much more pronounced than in Castlefield. Residents perceive the regeneration as disrupting their lives. Physical changes directly transform the practices afforded in the area as well as disrupting residents' attachments and 'sense of place'. The paradox between the users' discourses supporting the official strategies and their simultaneous resistance to changes stems from the area's desperate need to improve housing conditions and living conditions, so that initially any changes are welcomed.

Conclusion

Castlefield and El Raval are gradually becoming integrated into the spaces of flows of global tourism and investment and are regarded by their respective cities as examples of successful post-industrial planning. However, when going 'down below' and interrogating the perceptions of those engaging with these neighbourhoods on the ground, the cohesive narrative of regeneration becomes complicated and destabilized. The various users subvert hegemonic meanings by creating understandings of place sometimes related to, but at other times independent of, the representations offered in the officials' discourse. Resistance is shown here as 'not exactly an opposition, nor a separation of one space (of domination) from another (of resistance), but a *dis-located* interaction between the two, or three or more spaces. Resistance, then, not only takes place in place, but also seeks to appropriate space, to make new space' (Pile 1997: 16). Users produce competing narratives of place and thereby resist the cohesiveness and fixed identity that planners try to construct of places. The senses are a crucial ingredient in these narratives. Positive or negative experienced sensory perceptions are drawn upon to legitimate experiences of attachment, detachment or fear and to evaluate space and spatial relations. The visual hegemony of the imposed 'designer heritage aesthetic' and the restrictive themes and normative spatial practices of heritage, culture and leisure are disrupted by the embodied experiences of users that appropriate the same space in multiple ways, constructing a variety of stories, and illustrating how the sensuous body operates in Lefebvre's sense as a differential field. Understanding these gestures of place and sensory embodiments provide important clues on how power and resistance work side by side in everyday urban life.

While users clearly perceive a changed publicness in their neighbourhood as the economy of access has increased, this is a constrained access as social groups are excluded through forms and levels of place attachment, emotional identification and practices afforded by regenerated environments. Users have an ambivalent relation to the regeneration processes and at times question its legitimacy. However, users in both areas reproduce official representations when they support the need for more sanitized public spaces, invoking the officials' sensuous landscape of fear. In both places the actual needs of the locality are ignored and the regeneration acts more as a superimposed space, which does not aim to enter into a dialogue with the local context. The next chapter examines how the lived experience of space, as experienced through a sensuous 'total body', evades hegemonic experiences and practices.

8 Living in regenerated worlds

> The more carefully one examines space, considering it not only with the eyes, not only with the intellect, but also with the senses, with the total body, the more clearly one becomes aware of the conflicts at work within it, conflicts which foster the explosion of abstract space and the production of a space that is other.
>
> (Lefebvre 1991: 391)

To examine the 'socially embedded aesthetics' of places and to understand how they shape the politics of public life, one last aspect needs to be taken into account: the lived experience of place – in other words, the ways in which the senses define our direct experience in public life. I am following the lead of Soja (2000), who argues that to capture the complexity of real and imagined space – physical and representational space – and to examine space as the active arena of development and change, conflict and resistance, one needs to analyse lived space which is 'a simultaneously real and imagined, actual and virtual, locus of structured individual and collective experience and agency' (Soja 2000: 11). Let me briefly elaborate on this point. Lived space depends on the sensuous body to organize, mediate and 'make sense' of the spatial practices and mental constructs that produce space: 'spatial practice is lived directly before it is conceptualised' (Lefebvre 1991: 34). In Lefebvre's conception of space the body is central in that it mediates between mental, individual and social realms. Hence it brings together the subjective and the social, the public and the personal, the abstract and the concrete, the global and the local (Amin and Thrift 2002). This is an active and dynamic process, directly lived and mostly not reflected upon, an 'often untheorised understanding of space' (Shields 1991: 54). It is in lived space that contestations around meanings and uses of public space are played out, where spatial conflicts erupt and become tangible, where living together is negotiated.

To capture the often evasive dynamism of lived space, I start by exploring the activity rhythms of public spaces in Castlefield and El Raval and follow with an examination of the sensory mappings of these regenerated neighbourhoods: '[t]he rhythms of the city are the coordinates through which inhabitants and visitors frame and order urban experience' (Amin and Thift 2002: 17). Hence examining the activity and sensuous rhythms of an area provides us with an insight into the ways in which the senses underpin the social relations inscribed in landscapes, as different sensescapes fluctuate in their intensity and in their relationships. In the second half of this chapter I pay attention to the specific spatial contests that are emerging in these regenerated neighbourhoods. These conflicts around space provide an insight into the forms of publicness that are emerging in regenerated spaces. Focusing on the lived experience of places will bring us closer to the complexity of cityscapes in which different sensibilities both coexist and conflict.

Rhythms of everyday life

In Castlefield activity rhythms are shaped by outsiders. Its main streets become busy during rush hour in the early morning and early evening when streams of professionals drive to the various media businesses in the area and residents stride purposefully to work along the main thoroughfares into the city. The sharp clicking of female office workers' stiletto heels reverberates along the streets. In the daytime Castlefield is largely a subdued and quiet place. Delivery vans and office workers smoking outside their buildings are the only sign of activity. The only other people one encounters are likely to be tourists or groups of school children heading down Liverpool Street to visit the Museum of Science and Industry. Kate, an established resident, comments:

> In the daytime it's just dead really, it's people walking through going to work. I mean it is not a very functional place, not many people work in it. There's the design studio at the end and the engineering places but it is not a real 'work' area.

At lunchtime a burst of activity can be observed in the Urban Heritage Park when professionals from the city drive to Castlefield to have lunch in and around the bars offering European food, such as tapas or continental cheese and hams. Similarly, the fish and chip shop and Debby's café are packed as they '[provide] non-tourist food to people . . . who work in the area' (Jane, tourist guide), such as a full English breakfast, filled sandwiches and cups of tea and coffee. Activity resumes in the early evening when workers leave Castlefield and residents return. After 6 p.m. the area is mostly deserted. During weekends and school holidays activity increases dramatically. In the daytime families and a steadily rising number of foreign visitors come to visit the museums and the Heritage Park. In the summer, on bank holidays, a market opens on Liverpool Street which attracts thousands of people. Castlefield's subdued atmosphere then radically changes as one is immersed in a series of sensory engagements. The sensescapes of the market operate in a cooperative fashion, meaning that all the senses are equally engaged in the framing of place (Rodaway 1994). The jumbled display of goods is supported by a multiplicity of sounds such as street vendors loudly praising their goods and music blaring from the merry-go-round. There are the unexpected whiffs of smells such as a sharp odour of carpet-cleaner, the greasy smell of hot dogs. People's perspiration and perfumes are perceptible as they brush past each other in search of bargains. In Pete's view, a visitor, Castlefield then adopts:

> a holiday atmosphere. It's a bit like going to the coast, let's say in Wales. I mean the only water we've got is the ship canal water, and then you can go to the market and the fair so I suppose it reminds me of my holidays.

At weekends the bars on the other side of the Heritage Park attract a different kind of clientele, mainly young people with high disposable incomes. The atmosphere here is conceptualized by official agents and many of the residents as a 'European feel' (Laura). The sensescapes around the bar areas are more restrained and regulated than those by the market, leaving little room for the unexpected. They afford a predominantly visual and aural engagement, as terraces encourage panoramic views over the shimmering canals. In the Heritage Park people sit watching each other against a backing track of café music and echoing laughter. Castlefield's

Figure 8.1 The bank holiday market in Castlefield. (Photograph by author.)

afforded sensescapes, its open spaces, the water features, the 'designer heritage aesthetic' are central in attracting these customers to the bars:

> Barça and Dukes wouldn't be as popular if they didn't have the open spaces around them where people could sit during the summer. I mean, the space between the two bars belongs to them. It is packed in summer and on a sunny Sunday. It's a fact that it is one of the few pedestrian areas in the summer where you can sit out, there are not many pubs where you can sit out in the city.
>
> (Janet, established resident)

In summer, as well as on weekend nights, these bars and their surrounding public places are 'heaving', as young people parade up and down the bars in open-top cars with music blaring. One bar manager describes it as 'posers' paradise' (Tom). The spatial layout of the bars' outdoor spaces, their integration within Castlefield's public space, affords the display and free movement of bodies reminiscent of Mediterranean resorts: 'it was like being in Ibiza where everybody was wearing hardly any clothes and all were dressed up and it was all outside. There was a lot of movement between [Barça and Dukes], as if people were parading around looking cool' (Kate, established resident).

In winter, activities are fewer and far between, concentrated primarily around celebrations on New Year's Eve. Residents and shop-owners in the

Figure 8.2 The 'European feel' in Castlefield: Dukes 92 bar. (Photograph by author.)

area contrast the atmosphere in winter ('bleak, quiet and tranquil') with the summertime ('lively, loud and busy'). These responses illustrate how Castlefield's seasonal and fragmented public life is reflected in its extremely variable sensuous rhythms.

If we consider how spatial activities are linked to the ethics of engagement in Castlefield, it becomes clear that encounters between different social groups are not exactly promoted by the environment. Diverse user groups – young and economically upwardly mobile people, families, and residents – tend to use different areas of Castlefield and take part in separate activities, engaging in contrasting sensescapes. Most visitors to bars that I interviewed for this research were unaware of Castlefield's identity as an Urban Heritage Park, did not know about the nearby bank holiday market, and identified it predominately as an eating and drinking venue. Similarly, people attending the market or events held in the Arena did not go to any of the bars in the area. This suggests that the rhythms of activity in Castlefield are marked by the predominance of one group of people in a particular segment of public space. This is because public spaces in Castlefield have been conceived as having specific functions; their affordances restrict what you can do and who uses these spaces. In the design of Castlefield's public space there is a distinct lack of threshold spaces through which different activities could merge. This is further emphasized by the careful management of spatial activities that contains them within particular areas. Hence there is no criss-crossing of activity rhythms in the area, but one dominating rhythm that determines and frames the experience of public space. Different social groups' temporal or spatial paths do not coincide, which creates a layered public life characterized by low ethics of engagement between different groups.

In contrast to the spatial practices in Castlefield, El Raval's activities are predominantly determined by those living in the area, as a new resident comments:

> You get a lot of noise. It is an active neighbourhood especially in the mornings . . . You already get the neighbour shouting at eight o'clock in the morning. And if it is summer you open the windows and you can hear it even more, people hang out outside the entrances of buildings.
>
> (Judith)

Early in the morning people living in the neighbourhood walk out of El Raval to take the metro and buses to work. Shortly afterwards employees of the new cultural institutions enter the area and students stroll to the university faculties through the main regenerated access routes. Between 9.30 and 10 a.m. the neighbourhood is filled with the rattling sound of opening shop shutters. Soon afterwards the first tourists venture into the spacious regenerated streets of El Raval, often stopping to sit down in the new street cafés. The narrow side streets are filled with women from a variety of ethnic backgrounds shopping, greeting each other, stopping at shop entrances to chat. It reflects El Raval's cohesive social life that has been developed over the years through the daily routines of its inhabitants. One cannot help but feel sensuously immersed in the local life of the neighbourhood as one is involved in the tactile ethics of engagement, sharing space on the narrow pavements with young and old, Catalan-Spanish, North Europeans, Moroccans, Pakistanis or Filipinos while trying to negotiate a path alongside beeping delivery vans squeezing their way through the narrow roads. In El Raval the public/private division of space gets blurred in the non-regenerated streets through the sensuous engagements, such as the merging of music, voices and hammering tools coming from shops and balconies, views into stores and window displays, open doors into workshops – the street becomes a domestic space. Alicia, a museum employee says that for her one of the distinctive features of El Raval is not only the large number of chance encounters one experiences but its mixture of smells:

> Through the street I walk there's a shop where they sell Arabic cakes, and that's a particular smell I notice, it's a very intense and sweet smell. Then there's a perfume shop which has also quite an intense smell and then of course you can also smell the rubbish on the streets. What grabs my attention is the mixture of smells from the shops.

Both the spatial and sensuous rhythms indicate a socially mixed use of public space and at first sight a strong ethics of engagement. At 9 p.m. El Raval begins to settle down as the shop shutters stop rattling. The streets gradually empty until eventually the aural and olfactory sensescapes of lived homes emanate on to the street. One can hear the subdued voices behind open windows, interspersed with the monotone of the TV; the smell of food being prepared saturates the air. Most of the time Raval's regenerated streets are characterized by a coordination of insider and outsider activity rhythms, meaning that different spatial practices constantly overlap temporally in its public spaces, so that during the daytime no rhythm dominates another.

This is not the case when staged events are organized on the Plaça dels Angels or on weekend nights when young people from all over Barcelona come to some of the new designer bars. Then the main regenerated access roads to the cultural quarter are filled with the noise of revellers until late at night. This has led to conflicts between residents and the city council, as residents have put banners on their balconies stating: 'We have a right to sleep' (see Degen 2004; Chatterton and Hollands 2003). This is a clear example of contestations over whose sensescapes are allowed to dominate public life.

However, the regeneration has not affected all areas with the same intensity and non-regenerated spaces remain mainly filled by local activities. As in Castlefield, one can discern a segregated ethics of engagement, with newcomers and tourists being attracted to the designer bars and cultural events promoted by the council. This contrasts with much of the activities and practices in Raval's non-regenerated backstreets, where many shops are closed because of lack of business and few people walk. The spatial economy of access has created a zoned spatialization of the area divided into regenerated and non-regenerated spaces.

To end this section on the spatial rhythms in the regenerated areas I want to briefly focus on the flagship regenerated squares: the Event Arena in Castlefield and the Plaça dels Angels in El Raval. These new squares almost operate as 'magnifying glasses' for the neighbourhoods' reconfigured public life and as places where the 'designer heritage aesthetic' is at its most perceptible. The most common practice is the visual capturing of these places. One can observe school groups or individuals in these squares drawing or photographing the new prestigious architecture, such as the MACBA in El Raval or the regenerated heritage features such as the Victorian railway viaducts in Castlefield (Figure 8.3). Dustmen or urban park rangers make sure that these global stage-sets are constantly cleansed of any rubbish or graffiti that could disturb an easy visual consumption of place.

Squares or adjacent landmarks are often chosen locations for professional film teams and photographers (Figures 8.4 and 8.5). In fact the Museum of Modern Art in El Raval features in ads for anything from cars to cold sore cream, and Castlefield is used as a background for television programmes such as *Top Gear*. These practices serve to add yet more

Figure 8.3 Tourists posing in front of the Museum of Contemporary Art in El Raval. (Photograph by author.)

support to my earlier contention about the predominance of the visual sense in the conception of a 'designer heritage aesthetic'. These new public places have been conceived as places to gaze at and are used for visual practices by visitors. They have become commodities for global spatial consumption and one could argue that visitor activities dominate the rhythm of these squares.

However, this is not the case as these spaces are linked to their surroundings and embedded in the daily rhythms of their neighbourhoods. Hence on most days the Event Arena in Castlefield is eerily empty. In summer organized outdoor events are based in and around the Event Arena to attract large crowds of families with children and to fill the area with activity for a day. Yet, as soon as the event is over, people rapidly leave the neighbourhood, as Castlefield is perceived as lacking places to go and spend time in:

> There has to be something on to attract people. They don't come here on Sunday normally very much. You see the occasional people wandering around the canal but they only come here if there's something on.
>
> (Laura, established resident)

Figure 8.4 Photo-shoots in El Raval. (Photograph by author.)

Figure 8.5 Photo-shoots in Castlefield. (Photograph by author.)

On the Plaça dels Angels life starts at about 11 a.m. On days when the sun bathes the square, an elderly contingent congregates along the MACBA's ramps to rest their shopping bags and chat, while reclining tourists top up their tans and locals seek respite from the sun in the shady recesses of the building (Figures 8.6 and 8.7). Dog owners walk their pets over the square and an ever increasing number of skate-boarders use the museum ramps.

After lunch, during Spanish siesta time, the square is deserted. It changes in the late afternoon when Plaça dels Angels becomes a key meeting space for children in the neighbourhood. Their mothers congregate according to their cultural backgrounds: the Filipinos on one side, the Pakistanis on the other, the Spanish in a third area. Children run screaming around the square; rap music sounds from a ghetto blaster while the neighbourhood's teenagers meet on the steps of the museum. In winter this activity is restricted to a few hours until 8 p.m. In the summer the chatter and music can linger until midnight as the vast square is filled with a fresh breeze. A new resident evokes the scene:

> the space in front of the museum, the square, there are always children playing football, basketball. Some people spend whole afternoons there, you can even see them having dinner there. For example the Filipinos in summer have dinner there and spend time there. I guess it is because their flats are small and hot and the square is big and lively as children can play there. And people meet up with their friends, at least the people from the neighbourhood.
>
> (Julia, new resident)

Figure 8.6 Teenagers appropriating space. (Photograph by author.)

Figure 8.7 Elderly pedestrians resting. (Photograph by author.)

Figure 8.8 Children playing at night on the Plaça dels Angels. (Photograph by author.)

In an active demonstration of appropriation, locals have renamed the square 'the square of the nations', symbolizing the importance of immigration on the character of the neighbourhood and indicating the racialization of space, a feature I discuss later in this chapter.

What do the spatial practices in regenerated spaces tell us? First they show how in the lived experience of place the hegemony of the visual sense is challenged and they reveal how '[t]he characteristic features [of a neighbourhood] are really temporal and rhythmical, not visual' (Lefebvre 1996: 223). Further it shows that as instances of 'dominated space' (Lefebvre 1991) the flagships of cultural globalization can become 'appropriated space' through the quotidian practices of different user groups. Both in Castlefield and El Raval regeneration is embedded and digested through local practices. This reappropriation produces new configurations of public space and life based on the specific ways global and local processes interlock in each neighbourhood and create what I would describe as a 'third space of interaction' between the local and the global with its own particular features and new social conflicts.

In Castlefield this 'third space of interaction' is characterized by a *commercial leisurization* of public space. Public life in this neighbourhood, as it was indeed conceived by official agents, depends on regularly formalized commercial animation. The outcome is that there are extreme fluctuations of activity in the public life of the neighbourhood. Outsiders determine the rhythms of the space. One activity, which may be commuting, bar-life or organized events, determines the overall activity rhythms of the area. In Lefebvre's (1996) words, 'linear rhythms' predominate, stemming from mechanical human activity. Different activities are restricted to particular parts of the neighbourhood. Hence public life in Castlefield is extremely cyclical and dependent on the provision of commercialized events. This creates an area where public life is experienced by different user groups as very variable and characterized by high levels of activity or by its absence, and which, as I will show below, creates conflicts between the different groups claiming public space.

El Raval's 'third space of interaction', on the other hand, is characterized by a *lived localization* of public space, which means that despite the official agents' attempts to 'homogenize' El Raval's public life with the rest of the city, local practices and activities constantly reappropriate regenerated public spaces. In El Raval the global flagship area has been integrated into the neighbourhood through the routine activities of its residents. Linear rhythms are interspersed here with more cyclical rhythms that stem from the regular use of space by locals. One could argue that the global spaces are 'normalized' in El Raval through its insider uses. Residents resist the representations of officials by recreating *their* neighbourhood in a 'theme park'.

Sensuous mappings

Activity rhythms are intricately linked to sensuous rhythms. As public life is punctuated and produced through activities we experience these through the senses. An unfolding sensory landscape is created through past and present activities. Activities create events, yet it is the analysis of the sensuous rhythms produced that provides us with an insight into how places are experienced in a more phenomenological way as 'presences', thus evoking those inexpressible qualities of place. As Lefebvre summarizes: 'the act of rhythmanalysis integrates these things – this wall, this table, these trees – in a dramatic becoming, in an ensemble full of meaning, transforming them no longer into diverse things, but into presences' (Lefebvre 2004: 23). The sensuous maps I present here should not be regarded as fixed or definitive accounts but as a reflection of a 'moment' in the social production of space which provides us with an insight into the politics of public life as sensuously produced at the time of the study. As the neighbourhoods change and transform, these maps will keep changing.

To capture the sensuous rhythms ingrained in the physical fabric of the neighbourhood and to develop a sensuous map in each area I have combined items from my own ethnographic sensory mapping of each place with those from my interviewees' experiences. During interviews I asked interviewees to describe the neighbourhoods by a smell, a colour, a sound and an object. Tables 8.1 and 8.2 provide a summary of the responses provided.

Table 8.1 Castlefield

Smells	*Sounds*	*Colours*	*Objects/forms*
No smell	Building work	Sandstone	Railway viaducts
Fresh smell	Trains	Brick colour	Arches
Neutral smell	No sound	Red	Contrasting shapes
Car fumes	Quietness	Metal grey	Contrast between curved and straight
Different smells in different areas	Music (evening)	Opaque	Superimposed shapes

Table 8.2 El Raval

Smells	*Sounds*	*Colours*	*Objects/forms*
Bad smell	Voices of people talking/yelling	Muted colours	A three-dimensional square
Mixture of smells depending on shops	Echoing traffic in the narrow streets	Grey, brown	A net
Urine	Building work	Outer areas: yellow orange, interior areas black and grey	Winding streets
Sewage	– Bulldozers	Green plants on the balconies	Bands marching through the streets
Clean washing	– Noise of moving people	Contrasting colours: blue sky, dark streets	Balconies
Food	Children screaming	Grey with sparks of colour	Pakistani corner shops
No smell in the north, unpleasant smells in the south	Sound of TV	Mixture of lively and muted colours	Twisted streets
Urine in the south, antiseptic smell in the north		Dirty colours	Balconies with washing hanging out
Dampness			A labyrinth
			A box of surprises
			A womb

Let me briefly reflect what these descriptions tell us about the sensuous mapping of each neighbourhood. The common soundscape in both neighbourhoods during the time of the study was the sound of building work and bulldozing, signalling the process of physical transformation – but here is where the similarities end. When comparing the sensescapes of the two neighbourhoods, we can appreciate how Castlefield's respondents evoke a *spectacular place*, as architectural form and visual features dominate the description of Castlefield as an object: railways, arches, viaducts, shapes. Castlefield either lacks smells or is identified with fresh smells, reflecting the 'bland character' that residents evoked and the identification of car fume smells indicating its role as a thoroughfare to the city. The experience of different smells in different areas could be interpreted as a reflection of the segregated activities taking place. Quietness is the main defining feature of Castlefield, however interrupted by the sound of trains passing and the noise of bar visitors. The colourscapes refer to those mainly found in the

Urban Heritage Park: sandstone, brick colour, metal grey. The responses reflect a uniformity, a lack of contrast, in sensuous experiences which might indicate a sense of completeness of the regeneration. No reference is made to the sensescapes of the derelict warehouses that still exist in the non-regenerated areas. The sensescapes in Castlefield stand out for their lack of human life; nobody's presence is evoked in these descriptions. The sound of music coming out of bars in the evening seems to be the only sign of activity. One could claim that the 'commercial leisurization of public space' is reflected in Castlefield's sensuous mapping.

El Raval's characterization as a 'lived localization of public space' is similarly reflected in its sensuous mapping. The features evoked suggest that it is people's practices that produce the sounds, smells and tastescapes in El Raval: the voices of people, screams, urine, plants on balconies, blaring TV, and so forth. This reflects the contrasting space usage, and one could argue social demography, in both neighbourhoods. Castlefield's population of mainly under-45-year-olds are at work, whereas in El Raval many residents, especially women and the old, are at home. The sensescapes of El Raval afford corporeal experiences of the environment that are evoked through a range of metaphorical constructions, as if one object could not capture its multi-sensory nature such as a 'caja de sorpresas' (literally a box of surprises), a net, a womb, a labyrinth. Consequently, in contrast to Castlefield's *spectacular space*, El Raval can be described as a *dwelling space*. Strong contrasting experiences are evoked by all the senses in equal measure. Interviewees refer as much to the bad smells such as sewage as to the smell of food or clean washing. Most of the colour descriptions refer as much to the muted grey colour of the old buildings as to the sparks of colour produced by their inhabitants. Responses suggested that people in different areas within El Raval had very different experiences, highlighted by the difference that respondents make between the regenerated north (no smell or an antiseptic smell) and the non-regenerated south (urine or other unpleasant smells). These sensescapes are reflected in residents' meanings and associations with different areas in El Raval. Thus the northern part of the neighbourhood, with its new cultural venues and where much of the regeneration can be felt in the newly cleansed environment, is colloquially known as the 'cultivated' Raval. The south, where regeneration has focused more on improving social infrastructure and less in a renewal of space, is known as the 'oculto' (literally hidden, secret).

What my attempt at a sensuous mapping of neighbourhoods shows us is that the senses order and inform the spatial understanding of individuals' attachments to places. Not only have I started to indicate how one can try to measure it but I have also shown that there is a consensus of opinions and experiences about the sensuous geography of places which suggests a common sensuous imaginary. Last, I hope to have shown that each neighbourhood has a distinctive sensuous landscape which will frame and inform the experience of its public space and life. However, sensescapes are not stable, and rhythms intensify, alter or disappear as different social groups make their claims to space. In the next two sections I discuss the various forms of contestation that emerge in these newly designed environments. Focusing on each neighbourhood in turn, I start by analysing subversions linked to temporal activities in the neighbourhood and continue by examining specific instances of social conflict in each area.

Castlefield's 'taste wars'

What happens in these regenerated public places at night time? On Friday afternoons groups of young men and women begin to descend on Castlefield's string of bars. On winter weekends each of the bars can attract up to 800 people a night, swelling to 2000 in summer, when outdoor spaces are used as well. With six bars in the immediate Urban Heritage Park in 1999, this means that in summer the number of customers easily reaches 10,000. While during the day bar visitors are of mixed ages, at night the customers consist of 'the upwardly mobile 20–35s. Because it is the area to be seen in now' (Frank, bar manager). These bars are described negatively by many residents as 'boozer places' and are largely avoided. The sensescapes produced by the bar crowd – loud screaming, the stench of beer, empty bottles on the ground – are perceived as dominating and overpowering Castlefield's usually more sedate rhythms:

> I like living here. I think it's noisier than we expected. When we saw the plans for the development it looked more residential and quieter . . . There was one pub, which was very nice and didn't bother us, but the one opposite is noisy. It has speakers outside. More people hang around after the bars have closed . . . They are drunk and shriek . . . It seems to me to have become a kind of drinking centre.
>
> (Vanessa, established resident)

As becomes clear in interviews, residents feel that their sense of place is disrupted as they feel 'invaded' by activities and sensescapes produced by outsiders. What many of the interviews indicate between the lines is that residents regret the extreme promotion that Castlefield is getting as a place for night entertainment. As the number of visitors increases, and more and more public spaces are taken over by new bars, residents' sense of place ownership feels threatened:

> When we first moved in most of the drinking activity was on the other side of the canal. This gave us a certain feeling of separation. Now we have a bar this and that side, there's a restaurant opening actually in the building. There are far more people actually coming up and down this side and using what I suppose we feel is our space.
>
> (Vanessa, established resident)

While the officials' strategy of 'mixed use' has been successful economically in terms of attracting expensive residential development schemes and leisure outlets, the scheme has not been successful in accommodating the different needs of users in the area. Residents do not feel that they are living in a mixed-use space. For them 'it isn't mixed uses it is polarized' (Jim, established resident). Residents' feelings reflect the lack of opportunities given to them to add their concerns and views to the regeneration process. However, Shirley, an estate agent, argues that the market will provide a self-selection of residential needs as 'people won't buy in Manchester if they want peace and quiet. They are having to want to live in the area for the lifestyle really, and its hustle and bustle.' Castlefield's public life is increasingly oriented to a specific consumer group and lifestyle, so that residents who do not fit this prescribed lifestyle feel increasingly alienated from its spaces.

A different illustration of the emerging spatial contestations is the bank holiday market. Castlefield's Liverpool Road and access streets are closed off to traffic, and large numbers of visitors are attracted from the Greater Manchester conurbation into the area. Liverpool Road is then filled with stalls selling cheap clothing and food stalls, with vendors and customers haggling over goods. It is a racially mixed crowd, where many vendors seem to know each other as they lean chatting against the walls. My interviews with visitors reveal that many of them are regulars. However, they seldom visit Castlefield at other times. As Pam, a mother of two, explains, Castlefield for them stands for 'a day out, a shopping day sort of thing, for clothes and things'. Much of the attraction that Castlefield has for these visitors is the market atmosphere, being surrounded by other people, looking for bargains and going with the children to a fun-fair situated next to the Roman fort – activities that illustrate the 'compulsion of proximity' (Boden and Molotch 1992) and the sensuous enjoyment that participants gain from each other's presence. Pete, a market visitor, describes his first encounter with the market as a 'dream atmosphere', a description that emphasizes Castlefield's socially differential character within Manchester: 'About two or three years ago I was walking down Water Street, just turned round the corner and there was a big, big carnival atmosphere here . . . it was like a dream atmosphere with all this going on at a distance' (Pete, market visitor). Most of the market visitors stay on Liverpool Road and do not venture into the Urban Heritage Park. In fact, when made aware of it, most visitors did not know where it was or what it meant.

Residents complement officials' representations that the market does not fit into the quality regeneration scheme of Castlefield. They describe it as a 'rough-tough market' (Mike). They do not oppose a market per se, but the type of market and the consumption style that is promoted, which offends middle-class conceptions of 'good taste'. A resident explains:

> What I object to actually is not the fact that there is a street market, it is the fact that it is a tacky street market . . . I would like them to have an

> antique market, or sort of up-market market, rather than a cheap rubbish market.
>
> (Laura, established resident)

When questioned about their objections, residents construct a discourse of sensuous pollution framed in similar terms to their dislike of bar visitors. In this discourse 'outsiders' are portrayed as sensuously polluting Castlefield's ordered public space with the smell of cooking oil, the shouting of vendors, the rubbish left over after the event. Rob explains: 'You get people shouting through megaphones: buy [this and that] and you hear it twenty times an hour and they leave terrible litter' (Rob, established resident). Furthermore, the lack of 'distinctiveness' of the market's is regarded as a disadvantage, as it does not represent Castlefield's exclusive character – which differentiates it from the typical northern city. One established resident would like 'something that is unique in that respect rather than a replication of what people can get in Stockport, or Bolton' (Sarah). For residents the market, as opposed to the middle-class-geared events at the Event Arena which most of them attend, is not regarded as an 'amenity', as it does not conform to their leisure and taste values:

> It's one of those places where people just wander around because it is there and they're eating their hamburgers and their candy floss, I mean it has no value in my view. It's not even classed as an amenity in that respect.
>
> 
> (Sarah, established resident)

Figure 8.9 Bargains at Castlefield's bank holiday market. (Photograph by author.)

Both spatial conflicts bring to light the interrelationships between class and cultural capital, as discussed in Bourdieu's *Distinction* (1984), and the production of space. Thus, one could argue that the conflict described is about the clashing interests of different tastescapes. What residents in Castlefield object to is not visitors per se, but the type of visitors who are not regarded as fitting into the lifestyle or social image of the area. Thus, on the one hand, we have the popular taste of those enjoying events that involve a more direct engagement and participation, such as the market or night revellers. On the other, we have high cultural taste, based on a non-engaged appreciation of an event within the Arena where there is a clear separation between audience and stage, and which emphasizes a passive body. We could also interpret these contestations as representing the residents' fears of the sensescapes of the local 'northern city' reclaiming the sanitized spaces of regeneration; or as a fear of the outsider who with his/her overpowering practices transgresses the established spatial order. Similar to residents in gated communities, residents in Castlefield have bought themselves into an area that promised a certain lifestyle, entailing specific expectations and conceived ideas of what kind of public space and life should be fostered there. These conflicts around whose sensescapes and spatial practices predominate are thus informed by concerns around lifestyle and can be described as 'taste wars'. The bank holiday market was closed in 2000, and most of the bars in Castlefield – with the exception of Dukes 92, which has become a gastro pub – have now closed as other areas of the city become more fashionable.

El Raval's 'place wars'

What happens in El Raval after dark? The Plaça dels Angels is illuminated with blue-white lights, its architectural designer landmarks lit spectacularly against the night sky. It would seem to be 'visual imperialism' is at its zenith. Yet as soon as the galleries, bookshops and cafés close, and the cultural office workers leave their institutions, the 'undesirable' features of the marginal Raval progressively reclaim the place. The streets and the square become quiet and deserted. The homeless start preparing their cardboard sheets to sleep in the niches of the museum. North African men meet on street corners and local youths play football noisily on the square. Others roar past in cars with spoilers and on mopeds. These sounds echo in the square and take over the neighbourhood. Substance abusers, drunkards, prostitutes and Barcelona's destitute return to the streets. Each morning, new graffiti has appeared on the walls, sometimes only as paintings, sometimes with more political statements (Figure 8.10). Needless to say, these are quickly whitewashed over (Figure 8.11).

To counter these activities – the transgression of designer space – the city council is resorting to cultural animation. It tries to control the nights of El Raval by promoting late opening hours for galleries, museums, bars and

Figure 8.10 'Fed up with the P.E.R.I'. (Photograph by author.)

Figure 8.11 Graffiti being whitewashed. (Photograph by author.)

bookshops and organizes music events on the square and surrounding cafés. The idea appears to be that, by stimulating more civilized activities in the area, less space will be available for 'undesirable' activities. Similar to the way in which Zukin describes the increasingly controlled and 'civilized' Bryant Park in New York catering for the middle classes, one can argue that 'the cultural strategies chosen to revitalize [El Raval] carry with them the implication of controlling diversity while recreating a consumable vision of civility' (1995: 31). Similar to the situation in Castlefield, during events the Plaça dels Angels is filled with 'organized' noises and stage lights. Crowds of people sit relatively passively around the square, watching the performed event. Illuminated buildings serve as both a free backdrop and a constant reminder of the regenerated values. Public space is transformed into a passive performance space in which an artificial community is created, united by the spectacle. It is a safe and controlled sensuous experience.

Yet such regulating strategies are constantly subverted in their own idiosyncratic manner in El Raval, as an established resident told me with a smirk. He recounted the exhibition of designer kitchenware in a tent on the square. A security guard became puzzled by a family that had been watching for over an hour a television set that was supposed to show a five-minute advert. When the guard walked over, he realized that the family had changed the videotape to watch a Bruce Lee movie.

What the above examples of sensuous subversions illustrate, both in Castlefield and El Raval, is that despite the 'theme-parking', the places are appropriated by local practices and are used in ways other than the

Figure 8.12 A fashion show on the Plaça dels Angels. (Photograph by author.)

conceived images and practices projected by planners. What is produced is an increasingly temporally and spatially segmented public life. Marcuse and van Kempen (2000) have defined this phenomenon with the appropriate metaphor of 'the layered city', referring to a temporal layering of activities and groups in the same environment that meet but never engage. These layers emphasize the pre-existent geographical quartering of cities. They create an increasingly fragmented public life in which different cultural practices are hemmed in by their own boundaries, diminishing the space for spontaneous encounters.

The spatial contestations in El Raval revolve around sets of issues that are different from those in Castlefield. I indicated in the last chapter how the accounts of established residents construct El Raval as a neighbourhood with very strong social networks. This is also reflected in the social rhythms of the neighbourhood 'determined by the forms and alliances which human groups give to themselves' (Lefebvre 1996: 234). Older-established residents and shop-owners construct a sense of 'close-knit community' through an embodied re-membering of place (Stewart 1988). They recall the lost sensescapes of El Raval such as the smell of a fig tree next to the convent by the Plaça dels Angels, the sounds of marching bands through the streets, an evocative description of shops and shop-owners, most often deceased, and so on. These sensescapes are linked to a particular configuration of public space and life that is experienced as disappearing in the present as new social groups settle in the area. This is shown in the following childhood account of an 83-year-old resident:

> The streets were narrow, the houses nearly touched each other, the streets were so narrow. Everybody knew each other. Now I don't know anyone. The shop which I had, in which I lived and in which I was born is still there . . . It was a dairy and my dad constructed a house in front of it. We were all born in the dairy. Now there are gay people living there, they have their lives and don't bother anyone. The street hasn't been regenerated but the people have changed.
>
> (Aurora, established resident)

This type of story shows how established residents construct the past of the neighbourhood as an homogenous and settled community. However, since the nineteenth century El Raval has been a place of transition, a place defined by immigration and different social groups constantly transforming the public life of the neighbourhood with new spatial and sensuous practices.

While one could describe the above account as an idealized version of El Raval's past social life, interviews with younger residents and newcomers similarly construct El Raval's current forms of sociability as its distinctive identity and attraction. El Raval's long-time reputation as a marginal place has isolated it from the rest of the city and has preserved social forms that have disappeared in the 'modern city'. For example, in El Raval, as in other areas of the Old City, it is still common to have chairs in shops for customers to sit on, either to wait until it is their turn to be served or just to have a chat. Many old people, often living on their own, come daily into a shop at regular times to meet other customers and exchange a few words. In El Raval's 'lived localization of public space' local shops, often family-owned and reliant on a regular clientele, play an important role as 'third spaces' where social networks are developed. This leads most of the interviewees to define El Raval as 'a village within the city' (Figure 8.13).

Figure 8.13 A village within the city. (Photograph by author.)

It is precisely because of El Raval's strong social cohesion that the regeneration strategies are experienced, as discussed in the last chapter, as first and foremost disrupting the social life of the area. Both new and established residents resist the official strategies of equalizing or homogenizing El Raval's public life. They actively construct their neighbourhood as 'different', presenting it as a place with stronger forms of sociality than in the rest of the city. Hence, not surprisingly, established residents feel a strong bond with their neighbourhood, which they express by stories such as the following:

> El Raval is not as impersonal as other areas in Barcelona. There you can be living in a place and you don't even know who's living opposite you. Here this doesn't happen, here it is a bit like in a village. We have all known each other all our lives, I still have my childhood friends and they are still the same ones. This is the warmth, the cordiality, of this neighbourhood.
>
> (Javier, baker)

Newcomers as well as tourists, similarly describe the 'different' character of the neighbourhood compared to the rest of Barcelona by alluding to the sensuous features of its public places. The particular neighbourhood feel that makes it differ from other spaces in the city is palpable in the sensuous geography of the place: 'that is something I don't see everyday: the narrow streets, the dark passageways, lots of balconies' (Irish tourist). For newcomers the attraction of El Raval lies in its particular place gestures and sensuous rhythms inscribed in its living landscape, such as its individual old shops and social interactions that are nowadays considered as non-existent in the rest of the city (Figure 8.14). Both for newcomers and tourists the sensescapes of the area become part of the package of experience they consume when choosing to settle or wander around the Old City of Barcelona. Albert, a new resident, sums this up by stating, 'it has much more magnetism than other neighbourhoods', and concludes:

> It comes from the narrow streets, from the old shops. It isn't the same as a commercial neighbourhood . . . Here you can walk into an alley, it's like old times, you can sometimes go into some place and you are suddenly in the 1960s. And there are also hidden places that nobody knows about, with the typical old-fashioned bar. And I just find El Raval a really nice place but not well used and it hasn't been treated well.
>
> (Albert, new resident)

By describing El Raval both as a positive place and yet a place that is 'not well used and hasn't been treated well' Albert sums up the contradictions of El Raval's social geography. The 'magnetism' that Albert refers to, El Raval's sensuous difference from the rest of the city, one that provides a variety of

Figure 8.14 An old stationery shop. (Photograph by author.)

temporalities and forms of socialization, has been produced by years of isolation, institutional neglect and lack of use by Barcelona's citizens. It became the recipient of all the desired but hidden vices of the city: 'In El Raval was everything that Barcelona wanted but did not want to see. Or in other words, everything it needed and did not want to have nearby: prostitution, drugs, bars' (Oscar, established resident). The irony is that in present times this marginality is one of the main attractions and consumption features for newcomers and visitors to the neighbourhood. In an ever more commercialized urban landscape they regard it as proof of 'authenticity', yet also something they can leave behind once they step out of its confines or close the doors of their regenerated apartments.

The main experience of 'change' that established residents of El Raval refer to does not so much concern the resignification of the area as a cultural quarter, as this is often experienced as a separate phenomenon running almost in parallel to their everyday life. More important for these residents is the resignification of the neighbourhood through non-European immigration. El Raval has the highest percentage of immigrants in any of Barcelona's districts. This unprecedented number of non-European immigration has taken politicians and policy-makers alike by surprise, indicating the divergence between expected outcomes of globalizing processes and unexpected ones.

The neighbourhood is divided into areas where different ethnic groups have settled: the north of El Raval belongs mainly to Filipino people, and the southern areas are divided up between Pakistani and Moroccan territory (see Subirats and Rius 2005 for a detailed breakdown). El Raval's main attraction for immigrants is its cheap rents. Despite rents having risen, many of the old flats are still sublet to large groups of immigrants, who at times take turns to sleep in these flats, bringing back practices from El Raval's most marginal days that were thought to have been eradicated. As the study by Subirats and Rius (2005) concludes, due to the stretching of social service provisions, El Raval is again experiencing similar figures of extreme poverty and marginality that it had at the time it was most densely inhabited in the 1950s.

Other attractions include the social support networks that make El Raval a space of centrality and a reference point for immigrant communities. Jahid, a member of a Moroccan association, explains:

> El Raval is multicultural and there is a good social network. That's why all the immigrants concentrate here because it is easier for them to integrate in these kinds of conditions, because friends give them information on how to find this or that.

Furthermore the spatial structure of El Raval offers recognizable sensescapes, reminding immigrants of their home cities: 'It is similar to the structures of living of the country of origin of many immigrants . . . Small neighbourhoods, narrow neighbourhoods, shops which are open till late, opportunities for small businesses' (Jahid). Immigrants are appropriating the neighbourhood by gradually inscribing their culture and uses in the landscape of El Raval. The halal butchers and Pakistani corner-shops that are open until late at night, internet cafés and call shops, mosques and prayer halls, are all visible markers of their contribution to a reordering of sensuous and activity rhythms and a quickly transforming spatial identity (Figure 8.15).

Figure 8.15 A Pakistani internet café and call shop. (Photograph by author.)

As some new residents argue, El Raval's 'new immigrants' are replacing the vanishing Catalan-Spanish street-life of El Raval. The gradual social change of the neighbourhood is sensuously experienced in the new languages that permeate El Raval's soundscapes, the new smellscapes emanating from its windows and 'the colourful saris that you can see everywhere' (Judith). Many newcomers state that their choice of residence is influenced by the multicultural character of the neighbourhood, which they state as the most characteristic feature of El Raval, rather than the cultural quarter:

> The most characteristic feature of El Raval is the mixture of races that are lived and experienced. The MACBA is a museum of modern art that they've recently made, it is OK and it is in El Raval, but it is not typical of the neighbourhood.
>
> (Albert, new resident)

Resident associations complement these discourses and official agents' representations of 'cultural richness' that immigration is providing, by stating: 'here we have a great plurality. Other neighbourhoods are much more uniform and therefore more boring. Here you get a lot of cultural richness' (Miriam, independent residents' association). Urban policies to deal with the challenges of multiculturalism have taken a while to emerge and seem to be always a step behind the lived reality on the streets. During interviews for this research in the late 1990s, immigration was a feature largely ignored by official agents and regarded as a passing trend. Indicative of this attitude was that even in 2000 immigrant associations or groups were not invited to El Raval's annual street party,[1] chaired by Barcelona's mayor and described ironically by the local paper as the 'mono-colour neighbourhood party' (*La Vanguardia*, 16 July 2000). While it is outside the remit of this book to discuss in detail the changes that non-European immigration has produced in Barcelona's urban policies and economic and social development (although see Monnet 2002; Sole 2008), I briefly outline here how it has shaped the relations between the lived neighbourhood and the conceived Raval in marketing campaigns.

Official strategies have changed in recent years as immigration has proved to be an attractive marketing tool to promote El Raval as a 'multicultural quarter' (see Richardson 2004). Since 2001 immigrant associations have been invited to take part in the street party. In 2004 for the first time, the official poster reflected the multicultural character of the neighbourhood by depicting a person of African origin playing jazz. This is ironic, as black Africans are one of the smallest immigrant groups in El Raval. In this context, portraying a black person playing jazz could be interpreted as a safe 'consumable' image of the ethnic Other. Rapidly transforming ethnic areas such El Raval are increasingly packaged and branded under the notion of 'multiculturalism' in which visitors eagerly consume a relatively 'safe' ethnic sensescape of food, music or dance. Many of the visitors interviewed stated that while the 'multicultural' or 'cool' character of the neighbourhood was their main reason for venturing into El Raval, they did not really diverge from the clearly signalled tourist corridors in the neighbourhood. One could argue that while urban diversity is coopted as a form of cultural capital in promoting urban renewal and wealth (Keith 2005), intimate contact is resisted. While certain kinds of multiculturalism become manifest through the sensuous ordering of the spatial, others, such as the poor living conditions of many immigrants, or the rising numbers of foreign women in the prostitution trade, are kept at the margins of the visitor's experience.

Many established residents and shop-owners view immigration in El Raval critically. They tend to link the decrease of social cohesion in 'their' neighbourhood and the destruction of its affective structure with an increase of non-European immigration which started in the mid-1990s. The resentment of Catalan–Spanish residents has focused on 'immigrants' whom they regard as a threat to local businesses (e.g. by taking over small corner shops), and as competing for the scarce social housing, not to mention a principal source of crime in the area. Immigration, for many established residents, is a feature that is threatening their ownership of El Raval and leads somebody like Javier to conclude sarcastically: 'from a neighbourhood of prostitution we are now becoming a neighbourhood of immigration'[2] (Javier, baker).

Established residents and shop-owners make their dislike of immigrants explicit in interviews by referring to the changing sensescapes in the neighbourhood, as the following quote illustrates. People's awareness of a 'racist discourse' is so ingrained that they do not refer to bodily sensuous differences but create a more complex narrative that constructs the racialized Other as 'different' through their sensuous-spatial practices.

> The other night the light that was coming out of these new flats caught my eye. From most of them came a light that we consider normal in a living room, from a lamp. And there was one flat occupied by a Pakistani family and what drew my attention to it was that they had a simple neon light, fluorescent . . . Then, looking at it from outside it is horrible, you see pleasant ambient lights and then you suddenly see this neon light that crosses the whole room and no curtains, because they don't need curtains.
>
> (Gema, established resident)

Gema's words reflect how cultural difference is disguised as an aesthetic problem. Similarly to the discourse of official agents who believe that the social problems of El Raval can be improved through giving it a physical facelift, Gema portrays the racial problems in the area as a visual issue by arguing that immigrants do not fit into the new regenerated aesthetic of the area. No mention is given of the fact that only 15 years ago many of El Raval's autochthonous population were living under similar circumstances.

Immigrants are used here as the scapegoat for residents' responses to the uncertainty that global processes such as economic restructuring and movements of population bring about. The flow of global capital has transformed the socio-economic character of this locality in far more drastic and sudden ways than are evident within other areas of the city precisely because El Raval was neglected for so long. The racist discourse is a response to the rapid changes that impact on the local scale. Similar to the scapegoating of heroin, immigrants are perceived by many established residents as dismembering the place. These residents respond sensuously to neutral phenomena.

The conflicts reveal something important for the sensuous geography. They show that the closer you get into the households, the more you move away from official representations of the regeneration, and the more you can experience a real fear. Established Catalan-Spanish residents feel that their sense of belonging is being threatened, and this is experienced in the disintegration of reference points in the everyday sensuous landscape that place them in 'their' neighbourhood.

> It was like a village, everybody knew each other. If you had to do a favour, you did it. Everybody knew you, the people from the bean-shop knew you and the people from the stockfish shop, everybody. And you knew Juanita and Rosa and that was important. But that doesn't exist anymore. Not even the neighbours know each other anymore. All the neighbours that were once there aren't there anymore. Now there are Filipinos, Moroccans and Pakistanis, that is it. I live on the Calle Paloma, when on a Saturday afternoon I have time off work and I go out on the balcony I can hear only Filipinos, Catalans none, with this I'm telling you everything.
>
> (Maria, established resident)

But this is not the whole story. In the summer of 2000 neighbours (mainly Catalan-Spaniards and Pakistanis) organized themselves into groups and started patrolling the streets of El Raval, and the Old City in general, to demonstrate against the increase of mainly petty crime associated with Moroccan youths.[3] In September one of these patrols almost lynched two Algerian citizens in the Old City, bursting into a semi-derelict building inhabited by groups of Algerians, taking out their personal valuables and burning them (see *La Vanguardia*, 21 September 2000). Some residents' associations in El Raval countered the above outbursts of discontent with non-violent demonstrations claiming: 'Security yes, racism no.' Yet the Old City was appearing again in the papers as a marginal area or, in the journalists' words, 'a conflictual zone': 'Despite the increase of police and detentions, neighbours and tourists are still terrified at the proliferation of crime' (*La Vanguardia*, 22 February 2001).

These incidents point towards important transformations in contemporary public life. Here, public space no longer operates in the form of a 'public sphere', as a space of 'rational debate and open communication' (Sheller and Urry 2003: 109) which supports civil society and fosters solidarity and public participation. Instead an implosion of 'publicness' is at work in which civil life is reduced to independent communities that 'fight' for their right to place. In Castlefield this is expressed in the form of 'taste wars' and in El Raval in what I describe as 'place wars'. What is really at stake behind the racism and prejudices in both areas is the ownership of the neighbourhood. In both areas we can detect a strong spatial identification between long-term residents and the material landscape, and there is a real sense of spatial identity being under threat. Residents in Castlefield feel invaded by outsiders who only come temporarily to engage with their neighbourhood. In El Raval residents use immigration as a scapegoat for the changing patterns of sociability in their area. In both areas users draw on sensuous experiences to establish boundaries between themselves and the 'outsiders', whose practices are described through a language of defilement. In these places the fear of the Other is a product of globalization processes, resentment and prejudice. Bauman (1999) argues that one of the most sinister and painful outcomes that globalization has brought about can be summed up in the German word *Unsicherheit*, translated into English as uncertainty, insecurity or 'unsafety'. Bauman (1999) explains that while uncertainty and insecurity – internal feelings – cannot be directly tackled, 'unsafety' can be turned into a physical problem that can be contained, i.e. by scapegoating certain social groups. However, the ontological *Unsicherheit* is not resolved, but is constantly readjusted as new dangers loom. In Bauman's words:

> The snag is though, that while doing something effectively to cure or at least mitigate insecurity and uncertainty calls for united action, most measures undertaken under the banner of safety are divisive; they sow mutual suspicion, set people apart, prompt them to sniff enemies and conspirators behind every contention or dissent, and in the end make the loners yet more lonely than before.
>
> (1999: 5–6)

Since the summer of 2000, moods have certainly calmed down and no major racist incidents in El Raval have been reported in the press. In fact more recently in 2004, and in tune with El Raval's trendsetting multicultural character, business people have decided to celebrate El Raval's urban experience with a verb: *ravalejar*. Posters, T-shirts and postcards are now available with this new inscription in the main languages of El Raval's inhabitants (Figure 8.16). For some, *ravalejar* is an attitude. Others say it is a way of living and enjoying the neighbourhood. The more cynical say it is clever place marketing. What is certain is that one mainly finds the new slogan in the businesses lining the tourist corridors of El Raval.

Conclusion

The analysis of socially embedded aesthetics in the lived space in Castlefield and El Raval has shown how sensory perceptions and experiences are mobilized in diverse ways by different groups in the making of place and claims to space. Both activity and sensuous rhythms are important barometers for assessing how socio-spatial practices are (re)configuring public life. While the 'designer heritage aesthetic' appeals at times to some sensibilities, and at times the regeneration's spatial strategies impose particular meanings

Figure 8.16 Let's *ravalejar*. (Photograph by author.)

and spatial practices, these are embedded within the wider socio-spatial geography of the area.

Users react in complex and ambivalent ways to the regeneration strategies. They create parallel spatializations that are not directly opposed to the regeneration but which attach alternative meanings to it such as Castlefield becoming known as a drinking place rather than a Urban Heritage Park by a younger public. In El Raval, recent non-European immigration is radically shaping the emergence of new sensescapes, maybe more intensely than the regeneration process. Similarly, through their daily practices, users transform the regenerated environments and affordances to their own ends, rather than those envisaged. For example, people use the ramps of the MACBA for a diversity of activities from skateboarding to sitting or sleeping. We can conclude that rather than dominating the spatial order and life of the area, global processes such as regeneration are embedded in and lived through local practices. Indeed, by analysing how people experience regeneration on a day-to-day level and how such a process impacts on the public life of the area I have shown that regeneration strategies certainly produce new expressions of public life, though rarely those intended by planners and politicians. Castlefield has become an area with a very transitional and sporadic public life, dependent on organized events or commercial venues and hardly integrated into the public life of Manchester. Similarly El Raval, rather than becoming more integrated into Barcelona as planners expected, has become an area of deepening social contrast and inequality, palpable in the public spaces of the area. While the newly regenerated spaces are integrated into the local life of the area, a two-tier geography is slowly emerging with tourist or regeneration corridors next to streets of extreme poverty.

What do these examples of social conflict in El Raval's and Castlefield's public places tell us about the nature of public life? One can identify a clear layering of social worlds or lifeworlds that, although united through their geographical place experience, seem to live parallel existences. As Sennett (1990) laments about New York, this shows that while there is a clear sense of diversity in public places, an increased 'economy of access' involving a greater variety of social groups using the space, there is little interaction with the Other, fostering a poor ethics of engagement. While in both areas official agents claim to have created places that attract a more diverse public, my interviews and observations seem to indicate that these diverse publics could not be more distant from each other. My study has shown that the spatialization of these regenerated public spaces supports a segregation of activities and parallel public lives. The outcome is that public space as a political space of representation has imploded. The normative concept of public sphere as a space for negotiation has turned into a battlefield where different communities defend their right to space.

9 Conclusion

Regenerating public life?

> It is not, therefore, as though one had global (or conceived) space to one side and fragmented (or directly experienced) space on the other – rather as one might have an intact glass here and a broken glass or mirror over there. For space 'is' whole and broken, global and fractured, at one and the same time. Just as it is at once conceived, perceived, and directly lived.
>
> (Lefebvre 1991: 356)

When I started this book my aim was to understand more about the ways in which contemporary transformation of public space impinge on the public life of cities and how these physical changes are experienced by those involved in the 'making' of urban life, from planners and architects to residents and tourists. *Sensing Cities* has argued that examining the senses in daily urban life allows us to analyse the experience of public space and life as a political domain that links the personal lives of its inhabitants with broader structural changes in the city's politics and economy. The senses link the physical constitution of the city, its streets and buildings, with a lived social landscape. The senses are therefore central in the (re)configuration of public space and for shaping the politics of publicness. In this concluding discussion I first synthesize the key findings that have emerged from my study; I then outline what a sensuous approach adds to the study of urban life; and I end by outlining some suggestions for the planning and creation of future public places.

Regenerated public space and life in the city has become a crucial marketing tool to compete on a global catwalk for investment and tourism (Chapter 2). As I have shown, this restructuring of public space leads to a reformulation of public life through the manipulation of the sensuous geography. In this way social groups, place identities and spatial practices get shifted around, controlled, supported or erased. Influenced by the work of Lefebvre (1991, 2004) on the production of space, and his notion of the practico-sensory body, I suggest that one way to examine the relationship between the built environment and the social life of particular public places is to evaluate their publicness, which is grounded in three sensuous arenas.

First, 'the economy of access' is concerned with evaluating the physical and symbolic access provided by the layout and design of places. Second, 'the ethics of engagement' refers to the forms of interaction amongst strangers that are facilitated through the spatial set-up of public spaces. And, third, 'the politics of representation' relates to the ways in which public place acts as a space where different groups can represent their needs and struggles and therefore be acknowledged in society. As I have outlined throughout the empirical chapters, this publicness is produced and negotiated in all three moments of Lefebvre's production of space: the conceived, the perceived and the lived.

My two case studies empirically demonstrate the importance of what was theorized in Chapters 3 and 4 as 'socially embedded aesthetics' – the ways in which our daily embodied sensuous experience is shaped by social values – in organizing the publicness of particular spaces. The point to stress here is that sensuous experiences are far from natural, but are underpinned by social ideologies and are therefore always linked to relations of power in society. Sensing cities has been shown as a cultural ideological practice, both in the ways that sensing is framed by social values and hierarchies, and in that the material layout of cities is structured by sensory regimes.

In both areas examined I identified common strategies of standardization and control of urban life as the regeneration is aimed at a purification of the existing (non-regenerated) space through the spatial techniques of access and a 'designer heritage aesthetic' (Chapter 6). Dissenting views and voices are quickly silenced by a hegemonic discourse that portrays regeneration as physically cleansing, economically invigorating and socially beneficial to the city. Castlefield was turned into a leisure and consumption zone by formalizing a new public life via its perceptual *differentiation* from Manchester. The transformation of El Raval into a cultural consumption zone was based upon a dilution of its existing public life and its physical and social *homogenization* into the rest of the city. To clarify the nature of my concern, urban regeneration is put into practice through the organization of the senses, both in terms of the ways in which the 'problems' of particular places are sensuously defined and in terms of the proposed solution to these problems. Both interventions have produced a reduction of publicness by influencing through the organization of the sensescapes who can access, engage and be represented in public life. These processes may also be understood as the creation of new moral and civic landscapes organized around particular aesthetic and cultural values to attract a middle-class public.

Overall, the regeneration in the two case study areas can be judged as having been successful in achieving planners' short-term objectives (see Chapter 7). Castlefield was unanimously experienced by different social groups as a physically enhanced place and perceived in the public imagination as an exclusive high-status location. In El Raval the regeneration strategies have been successful in improving its reputation through its recoding as a cultural quarter in the north of the neighbourhood and thereby attracting

Barcelona's citizens and tourists into this once shunned area, albeit not in the numbers predicted. Yet my comparative analysis of the socially embedded aesthetics has also revealed that global processes, such as regeneration, are far from uniform and in fact impact differently upon separate local legacies as they are reworked through the specific histories and the socio-spatial practices of particular places (Chapter 5). While the regeneration schemes analysed here drew on similar spatial techniques to put into practice their cultural regeneration, and can be regarded as similar forms of spatial restructuring (Massey 1995a), the effects on place and people have been different.

This is highlighted through the complex and ambivalent ways in which different social groups reacted to the regeneration processes (Chapters 7 and 8). Certain users created parallel spatializations that, whilst not directly opposed to the regeneration, nevertheless attached alternative meanings to it. For example, in Castlefield, non-regenerated sensescapes were shown as important spaces for evoking memories and allowing room for fantasy outside normative encoded meanings. Some used their daily practices to transform the regenerated environment and affordances to their own ends, rather than those envisaged by planners. In El Raval, the museum slopes, rather than guiding crowds into the museum, are used for sitting, skateboarding, walking, break-dancing and so on. Yet, at other times, users would accommodate official discourses by reiterating the perception of non-regenerated areas as marginal and experiencing regeneration as a positive reversal of that marginalization.

Other users directly opposed the regeneration. In El Raval, for example, the aesthetic recoding was perceived as destroying significant perceptual markers of space. In this way a focus upon the senses during processes of urban change has exposed the deeper, existential significance of lived space, something which is often skated over superficially by devaluing it as an aesthetic concern and therefore a matter of taste. Regeneration is shown here as an embodied experience in which markers of recognition and emotional ties ingrained within the landscape are radically altered. The reconfiguration of sensescapes not only imposes a new organization of social life but also the transformation of a cultural landscape, a place of living memories. Focusing on the senses in the configuration of public life reveals an alternative geography of place by offering an insight into narratives, feelings, practices and experiences often hidden from common view. While I have hinted at the emotional landscapes that have been transformed through regeneration schemes it is clear that more work needs to be done in understanding the affective processes involved in place-making.

An examination of the lived experience in both areas in terms of activity and sensuous rhythms has shown that the interaction of local and global processes is developing new forms of spatial contest that threaten the social cohesion of the city. While these vary according to the particular characteristics, the history and the present condition of each local area, a common theme was that the new public spaces had broadened the economy of access of both neighbourhoods as more social groups were accessing the area. However, the spatial practices fostered by regeneration strategies also clearly emphasized a segregated temporal and spatial use of the areas, as public life came to be organized around transient collectives. Thus the ethics of engagement and the politics of representation were constrained in regenerated areas. Residents, for example, created purposive associations whose aims were not common civic goals but instead were instrumental, and territorially based. In Castlefield this took the form of 'taste wars', in which spatial contestations were related to differing lifestyles. By contrast, in El Raval the contests revolved around a 'right to dwell', or in other words 'place wars', and led to the scapegoating of immigrants for the perceived loss of sociability. In both cases residents constructed their discourses and place experiences around a threat of sensuous pollution to express their feelings of threat or dislike.

Examining regeneration as a set of sensuous transformations has revealed regeneration to be an ongoing, embodied and active process rather than merely a physical or economic project. The increasing pressure for urban competition to attract global investment has led 'official agents' in El Raval and Castlefield to adopt strategies aimed at fixing representations, framing interpretations, and controlling the lived experience of these places. Yet analysing local experiences uncovers a messier picture: a lack of coherence and regenerated areas as sites where a range of spatial practices and identities are articulated.

So how are we to judge these transformations in public spaces and life? As I have argued, the morphology of public spaces reflects and structures the social relations of a society. The senses then mediate these relations. With the rise of the entrepreneurial city, and with cities increasingly operating as global brands, we are witnessing the expansion of 'private space that pretend[s] to be public' (Hasenpflug cited in Klein 2000: 156). This means that one has to set aside the rhetoric that more 'public' access creates better public space. A 'good public' space must provide for the economy of access, but also ensure the ethics of engagement and the politics of representation. However, the ethics of engagement are not supported by the spatial configuration or by the spatial practices fostered in the regenerated public places analysed in this book. Public life in Castlefield is too transient and fragmented to encourage the mixing of different social groups. Moreover, access to and use of the neighbourhood is based on disposable income, making it difficult for people of lower-income groups to develop a sense of belonging or forms of spatial identification. In El Raval, there is a segregated ethics of engagements as the area becomes increasingly spatially and temporally divided into a 'bourgeois bohemian playground of cafés, galleries and boutiques . . . [and a] defiantly tough working class area that doubles as Barcelona's inner city red light district' (*Time Out* 2006: 104). In both neighbourhoods, existing social groups working and living there feel ignored or bypassed by regeneration strategies. Despite the regeneration process, the social problems of El Raval, such as poverty, prostitution and drug-dealing, still remain. In the case of Castlefield, both businesses in the area and residents feel that their needs come second to the promotion of the area for tourism and leisure.

With regard to the politics of representation the transformation of the sensescapes reveals a change in the meanings and presences fostered by the cultural landscape. The outcomes are ambivalent. On the one hand, one can identify an erasure of working-class histories and practices and a clear commercialization of public life. Moreover, my discussion of El Raval points towards privatization not only occurring when public squares are taken over by commercial events and gated by institutions, but also with regard to individuals' everyday practices, as the lack of balconies in new buildings makes residents turn inwards, away from the public life on the streets. The positive confusion of private and public in non-regenerated areas is increasingly threatened as the neighbourhood gradually becomes a place for tourists' visual appropriation, and regeneration processes try to draw a strict dividing line between public and private life.

Yet, looking at the recent developments in public space from a more positive perspective, one could argue that the centre for participation in these new public places is potentially stretched, as manifold 'mobile publics' are accessing these spaces and through their presence are shaping the everyday politics of the city. It is important not to idealize this process. As my discussion of the contestations in El Raval and Castlefield has illustrated, these are far from peaceful. Indeed a good public space will always leave room for struggle and contest, as Mitchell explains:

> what makes a space *public* – a space in which the cry and demand for the right to the city can be seen and heard – is not often its preordained 'publicness'. Rather, it is when, to fulfill a pressing need, some group or another *takes* space and through its actions *makes* it public.
>
> (Mitchell 2003: 35, emphasis in original)

The tension between 'local' and 'global', 'mobile' and 'settled' publics invites vital questions in terms of how to restore a more inclusive democracy in these new public spaces. The challenge posed by an increasingly mobile world is how to foster new forms of engagement in public places and participation in its public life that ideally includes a variety of attachments and patterns of spatial use. Due to lack of space I have not been able to work through how the politics of representation in public space are related to or inform issues of political representation in urban politics. Future studies need to assess the policy implications of the configurations of new public places and new forms of public life in terms of the rights and forms of citizenship that are fostered. For instance, the way in which immigrants have been largely ignored by the institutions in the reordering of El Raval's social and geographical space reveals the ethnocentric nature and provincialism of much urban planning. So, although references to multiculturalism are used to brand and sell certain areas of the city, in reality multicultural planning has yet to be embraced: 'the multicultural city needs to be conceptualized in part through a cartography that links the conventionally defined political institutions of governance and the state to arenas and spaces in which cultural politics regulates community life' (Keith 2005: 50). This research has only started to signal ethnicity as an important feature in the redefinition of public life in relation to regeneration, and more research is certainly required into the relationship between place, embodiment and 'ethnicity making' (Knowles 2003).

In part this study has been a celebration of urban life, but it is also intended as a contribution to recent studies that foreground the experiential dimensions in the city (see, for example, Pile 2005; Edensor 2005; Till 2005), and an attempt to capture some of the often inexpressible qualities of places. My argument throughout the book has been that only by getting close to the embodied sensuous experience can we understand the impact, or lack of impact, urban restructuring has upon social life in particular areas. Only by setting up this trialectic between body, space and the world, can we examine the variety of ways in which bodies make places and places make bodies.

So what do these findings and reflections tell us about the planning and creation of future public places? There are four main recommendations that I would like to reflect upon. First, this study shows the importance of consulting a multiplicity of users involved in the life of a neighbourhood, ranging from new and established residents to shop-owners, tourists and commuters, before any regeneration strategies are elaborated. The consultation would ideally involve the creation of a sensuous map of the area, based on activity patterns and sensory perceptions and overlaid onto a geographical map. This would offer various insights into the 'lived experience' of the neighbourhood. On the one hand it would show how the neighbourhood is spatially used and its emotional significance for different social groups; on the other it would indicate how the physical set-up supports and expresses the social life of an area.

The study has shown the significance of sensuous markers in providing a sense of continuity and belonging to city dwellers. People appreciate places of recognition, of emotional attachment and of surprise and discovery. Regenerated areas must offer a broad spectrum of opportunities to relate to the city and its broader public life, as 'it is only such sensuous signals, with lifelong validity, which make it possible at all to endure change' (Thomsen 1998: 125) – without limiting these efforts to rigid forms of communal belonging.

Second, each regeneration scheme must be conceived as an individual project, taking into account the historical and current social global context of the area. My comparative study has shown that specific lived patterns, such as particular forms of street life, cannot be easily distilled into predictable forms and exported, but need to be carefully assessed in relation to the existing patterns of street life and culture.

Third, planning involves a danger of broad-brush approaches and oversimplification. As this study has shown, Castlefield's artificial orchestration of activities through commercial or institutional means leads to a space that is experienced as having extremes in public life, being either almost abandoned or frantically hectic. El Raval's spatialization into almost two zones of public spaces leads to a layered and fragmented public life. The neighbourhood is divided into places attracting middle-class global cultural activities alongside those catering for El Raval's local inhabitants, though the latter are no less international. Areas should be developed by keeping a certain flexibility in their design which can allow for and stimulate urban disorder as the environment is folded and unfolded by different city dwellers (Sennett 1996b).

Finally, regeneration must not be left solely to developers. Rather, planners need to be given more powers and instruments with which to control the process. The neo-liberal laissez-faire politics in urban planning has had the effect, as Harvey powerfully suggests, of leaving 'the fate of cities almost entirely at the mercy of real estate developers and speculators, office builders and finance capital' (1996: 41). The consequence has been that commercial considerations overpower civic ones and elite groups determine the look and feel of urban living, thus excluding certain practices, social groups and meanings because they do not conform to the goals of economic development. Consensus must be sought through democratic consultation and by ensuring that a diversity of social groups have a 'right to the city'. As public spaces are the material location where we learn to live together, where an important part of our everyday socialization takes place, there must be space for 'a messy spontaneity of politics' (Mitchell 2003: 229) as a way to further the negotiation of different social positions. Without this, public life as such loses its meaning and political function.

Ultimately both planners and academics would benefit from stepping back from their 'grand designs' and 'getting close to other people, listening to them, making way for them' (Sibley 1995: 184). As my study has illustrated, cities are not just physical spaces or social constructs but rather lived spaces, full of contradictions and contestations. It is this feature, their lived nature, that requires closer attention.

Notes

1 Introduction: Sensing cities

1. I use the concept of 'sensuous' rather than sensory to capture the socio-embodied and relational nature of the senses in our experience of place.
2. Smith (1996) includes in his book a chapter comparing US gentrification patterns with three European cities – Paris, Amsterdam and Budapest – but largely explored through secondary sources.
3. Lefebvre (1991) points out the problem of labels such as 'users' and 'inhabitants' whose marginalization by spatial practice even extends to language, as these labels often have connotations of marginalization or less privilege. I use this label to refer to the multiple social groups that use and experience these places: long-term residents, new residents, tourists, commuters and so on.
4. Domosh (1998) highlights the fact that the hegemonic norms of white male public space in nineteenth-century New York were constantly transgressed through the 'polite politics' of women and African-Americans of both sexes.
5. The ethnographic fieldwork was conducted between 1998 and 2000. Since then I have made regular visits to both areas to keep myself informed about their development.
6. In total I conducted 55 interviews in El Raval and 46 in Castlefield.

2 Public life in late modernity

1. This moment is also often translated in English as 'representational spaces'. This study adheres to Soja's (1996) translation.
2. Economy refers here to the ways in which access is managed as a resource, hence facilitated or discouraged through diverse means such as the built environment, marketing or transport.
3. It is not my intention to 'idealise' the agora, as I am aware that Greece was at that time a slave society and not everyone interacted as equals in this space.
4. Tonkiss (2005) has recently described urban sociality as being based on an ethics of indifference. While I am sympathetic to her arguments, I want to highlight through the term 'engagement' the embodied experience of the Other which individuals are constantly managing by expressing care or indifference.
5. I refer here to Putnam's (1995) definition of civil society as people framing associations that are neither commercial nor industrial.
6. Rifkin (2000) states that in the United States the phenomenon of 'common interest developments' is rising at an unprecedented pace of 4–5000 new developments a year, with 12 per cent of Americans already living in gated communities.
7. These are 300 foot square apartments built in Manchester's inner city that are

being sold for £70,000. They are 'the world's first transformable single space living unit. They're designed as trendy second homes for people who want to "krash" after a hectic day at work . . . small rooms which can be transformed into a lounge, bedroom, office or dining room' (*Manchester Evening News*, 13 July 2001).

3 Sensing the city

1 There is an expanding literature on this important topic. It cannot be surveyed in entirety here, but useful starting sources are: Butler and Parr (1999); Gleeson (1998); Kitchin and Lysaght (2003); Knowles (2003).
2 This is not to say that all bodies are the same. They are gendered, classed, racialized and aged, but thinking of the body as a medium draws attention to the common experiential parameters of relating to the world through the body.

4 Sensuous powers

1 In his analysis of public spaces in the late Victorian British town, Croll (1999) describes the unpredictable appearance of 'unwelcome characters' (often associated with the poor) in middle-class areas, or the high street as mobile 'dark spaces', and refers to the difficulty of constructing a meaningful moral geography of the streets.
2 For a detailed account on postmodern insecurities see Ellin (1997a: 25–6).
3 A critique of 'defensible space' is offered by Jane Jacobs (1961) who notes that the only way to ensure safety is the heavy and constant use of places by people. Safety, in her interpretation, is linked to the 'sensory openness' of places and the possibility of exposure to strangers.
4 Davis (1998) points out that those most paranoid about crime (those in white suburbia) live in areas with the lowest crime rates; thus security becomes a concept sold to those who can afford it:

> 'Security' becomes a positional good defined by income access to private 'protective services' and membership in some hardened residential enclave or residential suburb. As a prestige symbol – and sometimes as the decisive borderline between the merely well-off and the 'truly rich' – 'security' has less to do with personal safety than with the degree of personal insulation, in residential, work, consumption and travel environments, from 'unsavoury' groups and individuals, even crowds in general.
>
> (Davis 1998: 224)

5 Castlefield and El Raval

1 In 1825 St Matthew's Church and a Sunday School were built in Castlefield: 'To some middle class observers, the "working and lower classes" living at the Castlefield end of Deansgate could only be viewed with alarm – with suspicion if not dread – a class who required civilising. Education was a priority, and St Matthew's soon had a Sunday School lower down Liverpool Rd' (Brumhead and Wyke 1989: 29).
2 Historically Barcelona had three city walls: a Roman wall surrounding the current Gothic quarter; then, as the city grew, a second wall constructed in the Middle Ages, leading up to today's Ramblas; and a third wall built in the fourteenth century that included El Raval within the city limits.
3 The destruction of the second wall and the construction of the Ramblas as a 'boulevard' only started in 1768.

4 At the beginning of the nineteenth century El Raval had 612 looms, and by 1829 this figure had grown to 2014 (Ayuntamiento Barcelona 1980).
5 For a detailed discussion on the creation of the 'Eixample' see Hall (1997).
6 For a discussion see Naylon (1981: 245).
7 The name Raval has only been revived recently. Since the eighteenth century until recently it has been known as Distrito V.
8 As Villar (1996) explains, the area acquired this name only in 1925 when the journalist Francisco Madrid published a number of articles called the 'Los bajos fondos de Barcelona' in which he mentioned for the first time the toponym Barrio Chino, apparently referring to San Francisco's infamous Chinese quarter.
9 In its heydays El Raval attracted many writers and journalists. Its atmosphere was evoked in a number of novels (for a detailed discussion see Carreras 1988), most famously Jean Genet's *Journal du voleur* ('Diary of a Thief') (1949) and Andre Pieyre de Mandiargues's *The Margin* (1969).
10 Villar (1996: 26) evokes the terrible conditions: 'In the big halls or in humid rooms, with the walls full of saltpetre, without light nor ventilation; on straw mattresses so close to each other that they looked like a single bed or on extended hay, were sleeping piled up men and women, children and old people.'
11 In 1991 the lower zone of Las Ramblas still provided 60 per cent of the city's prostitution (Gabancho 1991).
12 The years between 1970 and 1986 experienced the greatest decrease in population, which led to the lowest number of residents in a hundred years. In 1980 the Old City was one of the least populated districts in the city, with only 6.6 per cent of the population (Ajuntamiento de Barcelona 1989).
13 See, for example, Lowry for Manchester and Guesden and Colom for El Raval.
14 It was local enthusiasts in 1978 that persuaded the Greater Manchester Council to purchase the various local railway properties from British Rail such as the Central Station (Walker 1993).
15 For a detailed evaluation of Docklands regeneration see Foster (1999).
16 The number of Manchester city centre hotels has dramatically increased in recent years. Whilst in the 1980s the city centre had seven hotels, in 1990 it had 13 and by 2004 it had 35 (Williams 2003).
17 Design is defined as:

> the relationship between different buildings; the relationship between buildings and the streets, squares, parks, waterways and spaces which make up the public domain itself; the relationship of one part of a village, town or city with other parts; and the patterns of movement and activity that are thereby established: in short, the complex relationships between all the elements of built and unbuilt space.
>
> (Manchester City Council 1997c: 7)

18 GATPAC stands for Grupo d'Arquitectes i Tecnics Catalans per al Progress De l'Arquitectura Contemporanea.
19 Barcelona's city planning is shaped by a history of using cultural events to develop its urban landscape and attract public funding. In 1888 and 1929 it hosted the World Expositions, and in 1922 and 1926 it attempted to secure the Olympic Games, which all led to urban development in different areas of the city. The same strategy was employed after the 1992 Olympics, with Barcelona's successful bid to hold the 'Forum of Cultures' in 2004. This provided for the redevelopment of the western part of the city next to the sea, the industrial and working class area of Poblenou, into a new residential and cultural district.
20 Since 1979 Barcelona's city and metropolitan planning has been guided by the General Metropolitan Plan of Barcelona (PGMB). The PGMB meant a new

territorial order that created a clear regulation of public spaces, especially of transport and communications systems.

21 See Castells (1983) for the role of neighbourhood associations in Madrid's reconstruction.

22 The municipal area of Ciutat Vella (Old City) comprises four areas: El Raval, Barceloneta, El Gotic and Casc Antic.

23 PERIs are legal instruments that regulate specific urban areas and which are equivalent to partial plans, usually affecting complete neighbourhoods.

24 In 1980 only 3 per cent of the housing had all basic services (gas, water, electricity, heating, bathroom, toilet).

25 The ARI is constituted by:

- Promocio Ciutat Vella S.A.
- Institut Catala del Sol (Catalan Institute of Land)
- Ayuntamiento (City Council)
- Generalitat (Catalan Government).

26 By 2004 45.5 per cent of the buildings in El Raval had been renovated (Subirats and Rius 2005).

27 This policy did not allow the renewal of licences for sex or leisure businesses on streets narrower than five metres (thus it affected most streets of the southern Raval), since ambulances were unable enter the area. Furthermore, only certain areas were permitted to house bars or sex and leisure establishments. New sex businesses had to be located within a predetermined distance.

6 Planning regeneration

1 See Wilson and Grammenos (2005) on Chicago.

2 I deliberately draw here on Goffman's (1959) notion of social life being understood from a theatrical perspective in terms of understanding back-stages as the places where performances are routinely prepared and front and front-stages as the places where performances are presented.

3 See also Slater (2005) for a discussion of social mix and gentrification in Toronto.

7 Perceptions from 'down below'

1 As Villar states in his history of the neighbourhood, marijuana was the most popular drug in Barcelona until 1973, when heroin took over and changed the face of the Old City, and in particular of El Raval. As he explains: 'With the expansion of the heroin market the Old City suffered a radical transformation: dealers, junkies, crime and insecurity invaded its old arteries. The Barrio Chino has now definitely lost its frontiers, and that characteristic taste of literary glory, now almost imperceptible, was buried forever' (1996: 232).

2 A number of independent residents' groups are emerging in El Raval as they become disillusioned with the 'official residents' association', which is perceived as working too closely with the regeneration bodies.

3 Spaniards generally make a difference between 'extranjeros' (foreigners), meaning people from Western countries, and 'inmigrantes' (immigrants), meaning people from non-Western countries.

4 Before the change in rent laws in 1994, rents were protected, and permitted three generations to renew their rental contracts with no increase in rent.

5 *La Veu del Carrer* is the city journal produced by the Federation of Neighbourhood Associations, in which critical views on Barcelona's governance and

urban development patterns have been published. Among the contributors are social scientists and planners.

6 In *The Country and the City in the Modern Novel* Raymond Williams evokes how modern cities are characterized by the light they emit: 'The lights of the city. I go out in the dark, before bed, and look at that glow in the sky' (1975: 5).

7 Graves in Spain are broad concrete blocks with deep recesses in a wall used as a tomb housing about 80 bodies in a block, thus looking very much like tower blocks.

8 Living in regenerated worlds

1 In Spain each village or neighbourhood in a city holds an annual 'fiesta major' (grand party) usually on the local saint's day. This party can last between three days and a week and is celebrated with events and dinners on streets and squares. It plays an important role in the neighbourhood's community life.

2 It is important to point out here that while five years ago the rich districts of Barcelona like Sarria or St Gervasi had the highest number of foreigners (mainly EU residents, Japanese and American citizens), nobody described them as 'immigrant neighbourhoods'.

3 The rates of victimization (citizens declaring that they have been the victims of aggression) in the Old City have fallen in recent years while perceptions of security in the zone have risen.

Bibliography

Abella, M. (2004) *Ciutat Vella, El centre històric reviscolat*, Barcelona: Aula Barcelona.

Abu Lughold, J. L. (1999) *New York, Chicago, Los Angeles: America's Global Cities*, Minneapolis: University of Minnesota Press.

Ajuntament de Barcelona (1980) *Informe sociologico del Distrito V*, Barcelona: Area de Serveis Socials.

—— (1983) *Plan especial de reforma interior: El Raval*, Barcelona: Ajuntament de Barcelona.

—— (1989) *Primeres jornades Ciutat Vella*, Barcelona: Ajuntament de Barcelona.

—— (1991) *Segones jornades Ciutat Vella*, Barcelona: Ajuntament de Barcelona.

—— (1995) *Terceres jornades Ciutat Vella*, Barcelona: Ajuntament de Barcelona.

—— (1996) *La segona renovacio*, Barcelona: Sector de Urbanisme.

—— (2007) *La poblacio estrangera a Barcelona*, Barcelona: Informes Estadistics.

Akkar, M. (2005) 'The Changing "Publicness" of Contemporary Public Spaces: A Case Study of the Grey's Monument Area, Newcastle upon Tyne', *Urban Design International*, 10, 95–113.

Albertsen, N. (2000) 'The Artwork in the Semiosphere of Gestures: Wittgenstein, Gesture and Secondary Meaning', in J. Bakacsy, A. V. Munch and A. L. Sommer (eds) *Architecture Language Critique: Around Paul Engelman*, Amsterdam: Editions Rodopi.

Allen, J. (1999) 'Worlds within Cities', in D. Massey, J. Allen and S. Pile (eds) *City Worlds*, London: Open University.

—— (2000) 'Power: Its Institutional Guises (and Disguises)', in G. Hughes and R. Ferguson (eds) *Ordering Lives: Family, Work and Welfare*, London: Open University.

—— (2003) *Lost Geographies of Power*, Oxford: Blackwell.

—— (2006) 'Ambient Power: Berlin's Potsdamer Platz and the Seductive Logic of Public Spaces', *Urban Studies*, 43 (2), 441–55.

Amin, A. and Thrift, N. (2002) *Re-Imagining the City*, Cambridge: Polity Press.

Atkinson, D. and Laurier, E. (1998) 'A Sanitized City? Social Exclusion at Bristol's 1996 International Festival of the Sea', *Geoforum*, 29 (2), 199–206.

Atkinson, R. and Bridge, G. (2005) *Gentrification in a Global Context*, London: Routledge.

Balibrea, M. P. (2001) 'Urbanism, Culture and the Postindustrial City: Challenging the Barcelona Model', *Journal of Spanish Cultural Studies*, 2 (2), 187–210.

—— (2004) 'Barcelona: del modelo a la marca', *Desacuerdos*: available at www.desacuerdos.org (accessed 26 June 2006).

Barthes, R. (1986) 'Semiology and the Urban', in M. Gottdiener and A. P. Lagopoulos (eds) *The City and the Sign: An Introduction to Urban Semiotics*, New York: Columbia University Press.

Bartolucci, M. (1996) 'Barcelona, the Current Construction Slowdown', *Metropolis*, 16 (1), 60–97.

Bauman, Z. (1993) 'The Sweet Scent of Decomposition', in C. Rojek and B. S. Turner (eds) *Forget Baudrillard?* London: Routledge.

—— (1999) *In Search of Politics*, Cambridge: Polity Press.

—— (2000) *Liquid Modernity*, Cambridge: Polity Press.

Benjamin, W. (1975) *Illuminations*, London: Fontana.

—— (1997) 'On some Motifs in Baudelaire', in N. Leach (ed.) *Rethinking Architecture*, London: Routledge.

Berger, J. (1972) *Ways of Seeing*, London: BBC.

Berman, M. (1983) *All that Is Solid Melts into Air*, New York: Verso.

Bianchini, F. and Parkinson, M. (1993) *Cultural Policy and Urban Regeneration*, Manchester: Manchester University Press.

Boddy, T. (1992) 'Underground and Overhead: Building the Analogous City', in M. Sorkin (ed.) *Variations on a Theme Park*, New York: Noonday Press.

Boden, D. and Molotch, H. (1994) 'The Compulsion of Proximity', in R. Friedland and D. Boden (eds) *Space, Time and Modernity,* Berkeley, CA: University of California Press.

Bohigas, O. (1986) *Reconstruccio de Barcelona*, Barcelona: Edicions 62.

Borden, I. (2000) 'Material Senses: Jaques Tati and Modern Architecture', *Architectural Design: Architecture and Film II*, 17 (1), 26–31.

Bourdieu, P. (1984) *Distinction: A Social Critique of the Judgement of Taste*, London: Routledge and Kegan Paul.

Bourdieu, P. (1999) *The Weight of the World*, Cambridge: Polity Press.

Boyer, M. C. (1988) 'The Return of Aesthetics to City Planning', *Society*, 25 (4), 49–56.

—— (1992) 'Cities for Sale: Merchandising History at South Street Seaport', in M. Sorkin (ed.) *Variations on a Theme Park*, New York: Noonday Press.

—— (1993) 'The City of Illusion: New York's Public Places', in P. Knox (ed.) *The Restless Urban Landscape*, Englewood Cliffs, NJ: Prentice Hall.

—— (1995) 'The Great Frame-Up: Fantastic Appearances in Contemporary Spatial Politics', in H. Liggett and D. Perry (eds) *Spatial Practices*, London: Sage.

Brady, E. (2001) 'Sniffing and Savouring: The Aesthetics of Smells and Tastes', in A. Light (ed.) *The Aesthetics of Everyday Life*, New York: Seven Bridges Press.

Brenner, N. (2004) *New State Spaces*, Oxford: Oxford University Press.

Brewer, J. (1997) *The Pleasures of the Imagination: English Culture in the Eighteenth Century*, London: HarperCollins.

Bridge, G. and Watson, S. (2000) 'City Publics', in G. Bridge and S. Watson (eds) *A Companion to the City*, Oxford: Blackwell.

Brill, M. (1989) 'Transformation, Nostalgia and Illusion in Public Life and Public Places', in J. Altman and E. Zube (eds) *Public Places and Spaces*, New York: Plenum Press.

Brown, M. P. (1997) *RePlacing Citizenship: Aids Activism and Radical Democracy*, London: Guilford Press.

Brumhead, D. and Wyke, T. (1989) *A Walk Round Castlefield*, Manchester: Manchester Polytechnic.

Buck, N., Gordon, I., Harding, A. and Turok, I. (2005) *Changing Cities: Rethinking Urban Competitivness, Cohesion and Governance*, Basingstoke: Palgrave Macmillan.

Bull, M. (2000) *Sounding out the City: Personal Stereos and the Management of Everyday Life*, Oxford: Berg.

Bull, M., Gilroy, P., Howes, D. and Kahn, D. (2006) 'Introducing Sensory Studies', *The Senses and Society*, 1 (1), 5–7.

Burawoy, M. (2000) *Global Ethnography: Forces, Connections, and Imaginations in a Postmodern World*, Berkeley, CA: University of California Press.

Butler, R. and Parr, H. (1999) *Mind and Body Spaces: Geographies of Illness, Impairment and Disability*, London: Routledge.

Carr, S., Francis, M., Rivlin, L. and Stone, A. (1992) *Public Space*, Cambridge: Cambridge University Press.

Carreras, C. (1988) 'Paisaje urbano y novela', *Estudios Geograficos*, 191, 165–87.

Carreras i Verdaguer, C. (1993) 'Barcelona 92, una politica urbana tradicional', *Estudios Geograficos*, 212, 467–81.

Castells, M. (1983) *The City and the Grassroots*, London: Edward Arnold.

—— (1996) *The Rise of the Network Society*, Oxford: Blackwell.

—— (2002) 'The Culture of Cities in the Information Age', in I. Susser (ed.) *The Castells Reader on Cities and Social Theory*, Oxford: Blackwell.

Castlefield Centre (1996) *Visitor Information*, Manchester: Castlefield Management Company.

Central Manchester Development Corporation (CMDC) (1996) *1988–1996: Eight Years of Achievement*, Manchester: Central Manchester Development Corporation.

Centre for Local Economic Strategies (1992) *Social Regeneration: Directions for Urban Policy in the 1990s*, Manchester: CLES Monitoring Project.

Certeau, M. de (1984) *The Practice of Everyday Life*, Berkeley, CA: University of California.

—— (1985) 'Practices of Space', in M. Blansky (ed.) *On Signs*, Oxford: Blackwell.

Certeau, M. de and Giard, L. (1998) 'Ghosts in the City', in M. de Certeau, L. Giard and P. Mayol (eds) *The Practice of Everyday Life*, vol. 2, Minneapolis, MN: University of Minnesota Press.

Chatterton, P. and Hollands, R. (2003) *Urban Nightscapes: Youth Cultures, Pleasure Spaces and Corporate Power*, London: Routledge.

Classen, C. (1998) *The Colour of Angels*, London: Routledge.

Classen, C., Howes, D. and Synnott, A. (1994) *Aroma: The Cultural History of Smell*, London: Routledge.

Claver-Lopez, N. (1999) 'Consecuencias sociales de la estrategia urbanistica municipal de la renovacion de Ciutat Vella', unpublished MA dissertation, University of Barcelona.

Clifford, J. (1986) 'Introduction: Partial Truths', in J. Clifford and G. Markus (eds) *Writing Culture: The Poetics and Politics of Ethnography*, Berkeley, CA: University of California Press.

Clotet, L (1981) 'Del Liceo al Seminari', in O. Bohigas (ed.) *Plans i projectes per a Barcelona 1981/82*, Barcelona: Area de Urbanisme, Ajuntament de Barcelona.

Cochrane, A., Peck, J. and Tickell, A. (1996) 'Manchester Plays Games: Exploring the Local Politics of Globalisation', *Urban Studies*, 33 (8), 1319–36.

Colom, J. (1999) *El Carrer*, Barcelona: Museu Nacional d'Art de Catalunya.

Corbin, A. (1986) *The Foul and the Fragrant*, Leamington Spa: Berg.

Cresswell, T. (1996) *In Place/Out of Place*, Minneapolis, MN: Minnesota Press.

Crilley, D. (1993) 'Megastructures and Urban Changes, Aesthetics, Ideology and Design', in P. Knox (ed.) *The Restless Urban Landscape*, Englewood Cliffs, NJ: Prentice Hall.

Croll, A. (1999) 'Street Disorder, Surveillance and Shame: Regulating Behaviour in the Public Spaces of the Late Victorian British Town', *Social History*, 24 (3), 251–68.

Crossley, N. (1996) 'Body-Subject/Body-Power: Agency, Inscription and Control in Foucault and Merleau-Ponty', *Body and Society*, 2 (2), 99–116.

Davis, M. (1998) [1990] *City of Quartz*, London: Verso.

—— (1999) *Ecology of Fear*, New York: Vintage Books.

Deansgate Quay Development Brochure (1999) Crosby Homes.

Dear, M. (2000) *The Postmodern Urban Condition*, Oxford: Blackwell.

Degen, M. (2004) 'Barcelona's Games: The Olympics, Urban Design and Global Tourism', in J. Urry and M. Sheller (eds) *Tourism Mobilities: Places to Play, Places in Play*, London: Routledge.

—— (2008) 'Modelling a New Barcelona: The Design of Public Life', in M. Garcia and M. Degen (eds) *The Meta-city: Barcelona – Transformation of a Metropolis*, Barcelona: Editorial Anthropos.

Degen M. and Hetherington, K. (2001) 'Hauntings', *Space and Culture*, 11/12, 1–6.

Degen, M. and Garcia, M. (2008) 'Barcelona – The Breakdown of a Virtuous Model?', *International Journal of Urban and Regional Research*, under review.

Degen, M., DeSilvey, C. and Rose, G. (2008) 'Experiencing Visualities in Designed Urban Environments: Learning from Milton Keynes', *Environment and Planning A*, in press.

Deleuze, G. (1992) 'Postscript on the Societies of Control', *October*, 59, 3–7.

Delgado, M. (1992) 'La cuidad mentirosa', *El Basílico*, 2ª época, 12, 16–23.

—— (1999) *El animal publico*, Barcelona: Anagrama.

—— (2005) *Elogi del Vianant: del 'model Barcelona' a la Barcelona real*, Barcelona: Edicions de 1984.

Derrida, J. (1973) *Speech and Phenomena and other Essays on Husserl's Theory of Signs*, Evanston, IL: Northwestern University Press.

Deutsche, R. (1996) *Evictions: Art and Spatial Politics*, Cambridge, MA: MIT Press.

Diken, B. (1998) *Strangers, Ambivalence and Social Theory*, Aldershot: Ashgate.

Domosh, M. (1998) 'Those "Gorgeous Incongruities": Polite Politics and Public Space on the Streets of Nineteenth-Century New York City', *Annals of the Association of American Geographers*, 88 (2), 209–26.

Douglas, M. (1966) *Purity and Danger*, London: Routledge and Kegan Paul.

Dovey, K. (1999) *Framing Places: Mediating Power in Built Form*, London: Routledge.

Duncan, N. (1996) 'Renegotiating Gender and Sexuality in Public and Private Spaces', in N. Duncan (ed.) *Body Space: Destabilizing Geographies of Gender and Sexuality*, London: Routledge.

Eade, J. (ed.) (1997) *Living the Global City*, London: Routledge.

Edensor, T. (1998) *Tourists at the Taj: Performance and Meaning at a Symbolic Site*, London: Routledge.

—— (2005) *Industrial Ruins: Space, Aesthetics and Materiality*, Oxford: Berg.

Ellin, N. (ed.) (1997a) *Architecture of Fear*, New York: Princeton Architectural Press.

—— (1997b) 'Shelter from the Storm or Form Follows Fear and Vice Versa', in N. Ellin (ed.) *Architecture of Fear*, New York: Princeton Architectural Press.

Evans, G. (2003) 'Hard Branding the Cultural City – From Prado to Prada', *International Journal of Urban and Regional Research*, 27 (2), 417–40.

Fainstein, S. and Judd, D. (1999) 'Global Forces, Local Strategies and Urban Tourism', in D. Judd and S. Fainstein (eds) *The Tourist City*, London: Yale University Press.

Featherstone, M. (1991) *Consumer Culture and Postmodernism*, London: Sage.

Featherstone M. and Frisby, D. (1997) *Simmel on Culture*, London: Sage.

Ferrer, M. and Caceres, R. (1992) *Barcelona espacio publico*, Barcelona: Ayuntamiento de Barcelona.

Ferrer, N. (1997) 'La nueva ciudad vieja', *CIC Construccion*, 309, 68–77.

Flusty, S. (1997) 'Building Paranoia', in N. Ellin (ed.) *Architecture of Fear*, New York: Princeton Architectural Press.

Foment Ciutat Vella (2003) *Report 2001–2002*, Barcelona, Ayuntament de Barcelona.

Font Arellano, A. (1991) 'Equipament collectiu i condicio residencial de Ciutat Vella', in *Segonas jornades de Ciutat Vella*, Barcelona: Ajuntamiento de Barcelona.

Fortuna, C. (2001) 'Soundscapes: The Sounding City and Urban Social Life', *Space and Culture*, 11/12, 70–86.

Foster, C. (1998) 'Texture: Old Material, Fresh Novelty', in A. Haapala (ed.) *The City as Cultural Metaphor*, Lahti, Finland: International Institute of Applied Aesthetics.

Foster, J. (1999) *Docklands: Cultures in Conflict*, London: UCL Press.

Foucault, M. (1977) *Discipline and Punish: The Birth of the Prison*, London: Penguin.

—— (1980) *Power/Knowledge*, Hemel Hempstead: Harvester.

—— (1986) 'Of Other Spaces', *Diacritics*, Spring, 22–7.

—— (1991) *The Foucault Reader*, ed. P. Rabinow, London: Penguin.

Fraser, N. (1990) 'Rethinking the Public Sphere: A Contribution to Actually Existing Democracy', *Social Text*, 25/26, 56–79.

Fyfe, N. R. (1998) *Images of the Street*, London: Routledge.

Gabancho, P. (1991) *El sol hi era alegra: la reforma urbanistica i social de Ciutat Vella*, Barcelona: Promocions Ciutat Vella S.A.

Garcia, M. (2006) 'Citizenship Practices and Urban Governance in European Cities', *Urban Studies*, 43 (4), 745–65.

Garcia, S. (1993) 'Barcelona und die olympischen Spiele', in H. Hausserman and W. Siebel (eds) *Festivalisierung der Stadtpolitik*, Opladen: Westdeutscher Verlag.

—— (1998) 'The Example of Barcelona's Inner City Area Ciutat Vella', in *The Hague: Upgrading the Route between Hollandsche Spoor Station and the Agora wit Spinplein/Nieuwe, Kerk/Bezemplein*, The Hague: Hague City Council.

Garcia-Ramon, M. and Albet, A. (2000) 'Commentary: Pre-Olympic and Post-Olympic Barcelona, a Model for Urban Regeneration Today?' *Environment and Planning A*, 32, 1331–4.

Genet, J. (1949) *Journal du voleur* ('Diary of a Thief'), Paris: Gallimard.
Gibson, J. (1986) *The Ecological Approach to Visual Perception*, Mahwah, NJ: Lawrence Erlbaum.
Gilloch, G. (1997) *Myth and Metropolis: Walter Benjamin and the City*, Polity Press: Cambridge.
Gleeson, G. (1998) *Geographies of Disability*, Routledge: London.
Goffman, E. (1959) *The Presentation of Self in Everyday Life*, London: Penguin.
Gomez, M. (1998) 'Reflective Images: The Case of Urban Regeneration in Glasgow and Bilbao', *International Journal of Urban and Regional Research*, 22, 106–21.
Gottdiener, M. and Lagopoulos, A. (1986) *The City and the Sign: An Introduction to Urban Semiotics*, New York: Columbia University Press.
Gordon, A. (1997) *Ghostly Matters: Haunting and the Sociological Imagination*, Minneapolis, MN: University of Minnesota Press.
Grigor, J. (1995) 'The Castlefield Renaissance', *Manchester Memoirs*: *Memoirs and Proceedings of the Manchester Literary and Philosophical Society*, 132, 61–70.
Grosz, E. (1998) 'Bodies-Cities', in H. Nast and S. Pile (eds) *Places through the Body*, London: Routledge.
Habermas, J. (1974) 'The Public Sphere', *New German Critique*, 3, 49–55.
Hall, E. (1969) *The Hidden Dimension*, London: Bodley Head.
Hall, T. (1997) *Planning Europe's Capital Cities*, London: Spon.
Hall, T. and Hubbard, P. (1998) *The Entrepreneurial City*, Chichester: Wiley.
Hands, D. and Parker, S. (2000) *Manchester: A Guide to Recent Architecture*, London: Ellipsis.
Hannigan, J. (1998) *Fantasy City*, London: Routledge.
Haraway, D. (1991) *Simians, Cyborgs and Women: The Reinvention of Nature*, London: Free Association Books.
Harvey, D. (1990) *The Condition of Postmodernity*, Oxford: Blackwell.
—— (1996) *Justice, Nature and the Geography of Difference*, Oxford: Blackwell.
—— (2000) *Spaces of Hope*, Edinburgh: Edinburgh University Press.
Hayden, D. (1996) 'The Power of Place: Urban Landscapes as Public History', in I. Borden (eds) *Strangely Familiar: Narratives of Architecture in the City*, London: Routledge.
Heaton, F. (1995) *The Manchester Village: Deansgate Remembered*, Manchester: N. Richardson.
Heeren, S. (2002) *La remodelacion de Ciutat Vella. Un analisis critico del modelo Barcelona*, Barcelona: Veins en Defensa de la Barcelona Vella.
Heidegger, M. (1977) 'Building, Dwelling, Thinking', in *The Questions Concerning Technology and Other Essays*, New York: Harper Torchbooks.
Herbert, S. (2000) 'For Ethnography', *Progress in Human Geography*, 24 (4), 550–68.
Hetherington, K. (1997) 'In Place of Geometry: The Materiality of Place', in K. Hetherington and R. Munro (eds) *Ideas of Difference*, Oxford: Blackwell.
Hetherington, P. (2006) 'Reach for the Sky', *The Guardian*, 7 June 2006.
Hill, D. M. (1994) *Citizens and Cities: Urban Policy in the 1990s*, London: Harvester Wheatsheaf.
Howes, D. (1991) 'Olfaction and Transition', in D. Howes (ed.) *The Varieties of Sensory Experience*, Toronto: University of Toronto Press.
—— (2005a) 'Hyperesthesia, or the Sensual Logic of Late Capitalism', in D. Howes (ed.) *The Empire of the Senses*, Oxford: Berg.

Howes, D. (2005b) *The Empire of the Senses*, Oxford: Berg.
Hubbard, P. (1996) 'Re-Imagining the City', *Geography*, 81 (1), 26–36.
—— (2004) 'Revenge and Injustice in the Revanchist City: Uncovering Masculinist Agendas', *Antipode*, 36 (4), 665–86.
Imrie, R. and Raco, M. (2003) *Urban Renaissance? New Labour, Community and Urban Policy*, Bristol: Policy Press.
Ingham, J., Purvis, M. and Clarke, D. B. (1999) 'Hearing Places, Making Spaces: Sonorous Geographies, Ephemeral Rhythms and the Blackburn Warehouse Parties', *Environment and Planning D*, 17, 283–305.
Jacobs, J. (1961) *The Death and Life of Great American Cities*, Harmondsworth: Pelican.
Jacobs, J. M. (1996) *Edge of Empire*, London: Routledge.
—— (1998) 'Staging Difference: Aestheticization and the Politics of Difference in Contemporary Cities', in R. Fischer and J. M. Jacobs (eds) *Cities of Difference*, London: Guildford Press.
Jameson, F. (1984) 'Postmodernism or the Cultural Logic of Late Capitalism', *New Left Review*, 146, 53–92.
Jay, M. (1993) *Downcast Eyes*, Berkeley, CA: University of California Press.
Jeffrey, P. and Pounder, J. (2000) 'Physical and Environmental Aspects', in P. Roberts and H. Sykes (eds) *Urban Regeneration*, London, Sage.
Jencks, C. (1996) 'The City that Never Sleeps', *The New Statesman*, 28 July, 26–8.
Jones, H. and Lansley, J. (1995) *Social Policy and the City*, Aldershot: Avebury.
Jones, L. (1997) *Manchester . . . The Sinister Side*, Nottingham: Wicked Publications.
Kearns, G. and Philo C. (1993) *Selling Places*, Oxford: Pergamon Press.
Keith, M. (2005) *After the Cosmopolitan? Multicultural Cities and the Future of Racism*, London: Routledge.
Keith, M. and Pile, S. (eds) (1993) *Place and the Politics of Identity*, London: Routledge.
Kilian, T. (1998) 'Public and Private, Power and Space', in A. Light and J. M. Smith (eds) *The Production of Public Space*, Oxford: Rowman and Littlefield.
King, A. (2004) *Spaces of Global Cultures: Architecture, Urbanism and Identity*, London: Routledge.
Kitchin, R. and Lysaght, K. (2003) 'Heterosexism and the Geographies of Everyday Life in Belfast, Northern Ireland', *Environment and Planning A*, 35, 489–510.
Klein, N. (2000) *No Logo*, London: Flamingo.
Knorr-Cetina, Karin (1997) 'Sociality with Objects', *Theory, Culture and Society*, 14 (4), 1–30.
Knowles, C. (2003) *Race and Social Analysis*, Sage: London.
Knox, P. L. (1991) 'The Restless Urban Landscape: Economic and Sociocultural Change and Transformation of Metropolitan Washington, DC', *Annals of the Association of American Geographers*, 81 (2), 181–209.
Lash, S. (1999) *Another Modernity: Another Rationality*, Oxford: Blackwell.
Lash, S. and Urry, J. (1994) *The Economies of Signs and Space*, London: Sage.
Latour, B. (1986) 'The Powers of Association', in J. Law (ed.) *Power, Action and Belief: A New Sociology of Knowledge*, London: Routledge and Kegan Paul.
—— (1993) *We Have Never Been Modern*, Hemel Hempstead: Harvester Wheatsheaf.
—— (2000) 'When Things Strike Back: A Possible Contribution of "Science Studies"', *British Journal of Sociology*, 51 (1), 107–23.

Law, J. (ed.) (1986) *Power, Action and Belief*, London: Routledge.
Law, L. (2001) 'Home Cooking: Filipino Women and Geographies of the Senses in Hong Kong', *Ecumene*, 8, 264–83.
—— (2005) 'Sensing the City: Urban Experiences', in P. Cloke, P. Crang and M. Goodwin (eds) *Introducing Human Geographies*, 2nd edn, London: Arnold.
Leach, N. (1997) *Rethinking Architecture.* London: Routledge.
Lees, L. (2003) 'Visions of "Urban Renaissance": The Urban Taskforce Report and the Urban White Paper', in R. Imrie and M. Raco (eds) *Urban Renaissance? New Labour, Community and Urban Policy*, Bristol: Policy Press.
Lefebvre, H. (1971) *Everyday Life in the Modern World*, London: Allen Lane.
—— (1976) *The Survival of Capitalism*, London: Allison and Busby.
—— (1991) *The Production of Space*, Oxford: Blackwell.
—— (1996) *Writings on Cities*, trans. E. Kofman and E. Lebas, Oxford: Blackwell.
—— (2004) *Rhythmanalysis*, London: Continuum.
Levinas, E. (1985) *Ethics and Infinity*, Pittsburgh, PA: Duquesne University Press.
Levine, M. V. (1987) 'Downtown Redevelopment as an Urban Growth Strategy: A Critical Appraisal of the Baltimore Renaissance', *Journal of Urban Affairs*, 9 (2), 103–23.
Ley, D. (1996) *The New Middle Class and the Remaking of the Central City*, London: Oxford University Press.
Lingis, A. (1994) *Foreign Bodies*, London: Routledge.
Lofland, L. (1973) *A World of Strangers*, New York: Basic Books.
Low, S. (1997) 'Urban Public Spaces as Representations of Culture', *Environment and Behaviour*, 29 (1), 3–33.
Lowe, D. (1982) *History of the Bourgeois Perception*, Brighton: Harvester.
Lukes, S. (1974) *Power: A Radical View*, London: Macmillan.
Lyon, D. (2001) *Surveillance Society*, Buckingham: Open University Press.
Lynch, K. (1976) 'Foreword', in G. T. Moore and R. G. Golledge (eds) *Environmental Knowing*, London: Hutchinson.
Madanipour, A. (2003) *Public and Private Space of the City*, London: Routledge.
—— (2006) 'Roles and Challenges of Urban Design', *Journal of Urban Design*, 11 (2), 173–93.
MacLeod, G. and Ward, K. (2002) 'Spaces of Utopia and Dystopia: Landscaping the Contemporary City', *Geografiska Annaler B*, 84 (3–4), 153–70.
McLuhan, M. (1962) *The Gutenberg Galaxy*, Toronto: University of Toronto Press.
McNeill, D. (1999) *Urban Change and the European Left*, London: Routledge.
Maffesoli, M. (1996) *The Time of the Tribes*, London: Sage.
Manchester City Council (1997a) *Castlefield: Strategy and Action Plan to Complete Regeneration*, Manchester: Manchester City Council.
—— (1997b) *City Development Guide*, Manchester: Manchester City Council.
—— (1997c) *The Manchester Plan First Monitoring Report*, Manchester: Manchester City Council.
Manchester City Planning Department (1980) *Castlefield*, Manchester: Manchester City Planning Department.
—— (1988) *Castlefield Conservation Area – Pamphlet*, Manchester: Manchester City Planning Department.
Marcuse, P. (1997) 'Walls of Fear and Walls of Support', in N. Ellin (ed.) *Architecture of Fear*, New York: Princeton Architectural Press.

Marcuse, P. and Kempen, R. van (2000) *Globalizing Cities: A New Spatial Order*, Oxford: Blackwell.

Markus, T. (1993) *Buildings and Power*, London: Routledge.

Marshall, T. (2000) 'Urban Planning and Governance: Is there a Barcelona Model?', *International Planning Studies*, 5 (3), 299–319.

—— (2004) *Transforming Barcelona*, London: Routledge.

Massey, D. (1995a) [1984] *Spatial Division: Social Relations and the Geography of Production*, London: Macmillan.

—— (1995b) 'The Conceptualisations of Place', in D. Massey and P. Jess (eds) *A Place in the World?*, Oxford: Oxford University Press.

—— (2005) *For Space*, London: Sage.

Mayol, P. (1998) 'The Neighbourhood', in M. de Certeau, L. Giard and P. Mayol (eds) *The Practice of Everyday Life*, vol. 2, Minneapolis, MN: University of Minnesota Press.

Melucci, A. (1989) *Nomads of the Present*, London: Hutchinson.

Merleau-Ponty, M. (1969) *The Essential Writings of Merleau-Ponty*, ed. A. Fisher, New York: Harcourt Brace and World.

Michael, M. and Still, A. (1992) 'A Resource for Resistance: Power-Knowledge and Affordance', *Theory and Society*, 21, 869–88.

Miles, M. (2000) *The Uses of Decoration*, Chichester: Wiley.

—— (2005) 'Interruptions: Testing the Rhetoric of Culturally Led Urban Development', *Urban Studies*, 42 (5/6), 889–911.

Miles, M. and Miles, S. (2004) *Consuming Cities*, London: Palgrave Macmillan.

Mitchell, D. (1995) 'The End of Public Space? People's Park, Definitions of Public, and Democracy', *Annals of the Association of American Geographers*, 85, 108–33.

—— (2003) *The Right to the City: Social Justice and the Fight for Public Space*, New York: Guildford Press.

Molotch, H. (1998) 'L.A. as Design Product: How Art Works in a Regional Economy', in A. Scott and E. Soja (eds) *The City*, Berkeley, CA: University of California Press.

Monnett, N. (2002) *La formacion del espacio publico*, Madrid: Catarata.

Mumford, L. (1961) *The City in History*, London: Secker and Warburg.

Nast, H. and Pile, S. (1998a) *Places through the Body*, London: Routledge.

—— (1998b) 'Introduction: MakingPlacesBodies', in H. Nast and S. Pile (eds) *Places through the Body*, London: Routledge.

Naylon, J. (1981) 'Barcelona', in M. Pacione (ed.) *Urban Problems and Planning in the Developed World*, London: Croom Helm.

Newman, O. (1972) *Defensible Space: Crime Prevention through Urban Design*, New York: Macmillan.

O'Connor, J. and Wynne, D. (1996) 'Left Loafing: City Cultures and Postmodern Lifestyles', in J. O'Connor and D. Wynne (eds) *From the Margins to the Centre: Cultural Production and Consumption in the Post-Industrial City*, Aldershot: Arena.

Office of the Deputy Prime Minister (ODPM) (1998) *The Impact of Urban Development Corporations in Leeds, Bristol and Central Manchester*, available at: www.odpm.gov.uk/index.asp?id = 1128645 (accessed 3 May 2007).

Office of National Statistics (2004) *Manchester City Centre Population Estimate*, available at: www.manchester.gov.uk/planning/studies (accessed 3 May 2007).

Officers Working Party (1982) *Castlefield Tourism Development Plan (1980–82)*, Manchester: Report of the Officers Working Party.

Ong, W. J. (1971) 'World as View and World as Event', in P. Shepard and D. McKinley (eds) *Environ/mental: Essays on the Planet as Home*, New York: Houghton.

Pallasmaa, J. (2005) *The Eyes of the Skin*, Chicester: Wiley.

Parson, D. (1993) 'The Search for a Centre: The Recomposition of Race, Class and Space in Los Angeles', *International Journal of Urban and Regional Research*, 17 (2), 232–40.

Peck, J. and Tickel, A. (2002) 'Neoliberalizing Space', *Antipode*, 34 (3) 380–404.

Pieyre de Mandiargues, A. (1969) *The Margin*, London: Calder and Boyars.

Pile, S. (1996) *The Body and the City*, London: Routledge.

—— (1997) 'Introduction: Opposition, Political Identities and Spaces of Resistance', in S. Pile and M. Keith (eds) *Geographies of Resistance*, London: Routledge.

—— (2005) *Real Cities*, Sage: London.

Porteous, J. D. (1985) 'Smellscapes', *Progress in Human Geography*, 9 (3), 356–78.

Postrel, V. (2004) *The Substance of Style: How the Rise of Aesthetic Value Is Remaking Commerce, Culture, and Consciousness*, New York: Harper Perennial.

Purwar, N. (2004) *Space Invaders: Race, Gender and Bodies out of Place*, Oxford: Berg.

Putnam, R. (1995) 'Bowling Alone: America's Declining Social Capital', *Journal of Democracy*, 6 (1), 65–78.

Quilley, S. (2000) 'Manchester First: From Municipal Socialism to the Entrepreneurial City', *International Journal of Urban and Regional Research*, 24 (3), 601–15.

Reichl, A. J. (1999) *Reconstructing Times Square: Politics and Culture in Urban Development*, Lawrence, KS: University Press of Kansas.

Richardson, P. (2004) 'West Side Story', *Condé Nast Traveller*, June.

Rifkin, J. (2000) *The Age of Access*, London: Penguin.

Roberts, P. (2000) 'The Evolution, Definition and Purpose of Urban Regeneration', in P. Roberts and H. Sykes (eds) *Urban Regeneration*, London: Sage.

Rodaway, P. (1994) *Sensuous Geographies*, London: Routledge.

Rojek, C. (1995) *Decentering Leisure*, London: Sage.

Rose, N. (1999) *Powers of Freedom*, Cambridge: Cambridge University Press.

Rushdie, S. (1988) *The Satanic Verses*, London: Viking.

Samuel, R. (1994) *Theaters of Memory*, London: Verso.

Sassen, S. (1991) *The Global City: New York, London, Tokyo*, Princeton, NJ: Princeton University Press.

—— (1998) *Globalisation and its Discontents*, New York: New Press.

—— (1999) 'Whose City Is It? Globalisation and the Formation of New Claims', in R. A. Beauregard and S. Body-Gendrot (eds) *The Urban Moment*, London: Sage.

—— (2000) 'New Frontiers Facing Urban Sociology at the Millennium', *British Journal of Sociology*, 51 (1), 143–59.

Sassen, S. and Roost, F. (1999) 'The City: Strategic Site for the Global Entertainment Industry', in D. Judd and S. Fainstein (eds) *The Tourist City*, London: Yale University Press.

Schafer, R. M. (1977) *The Tuning of the World*, New York: Alfred A. Knopf.

Schmidt, C. (1994) *The Role of Tourism in Urban Regeneration*, Manchester: Report for the Centre of Environmental Interpretation.
Scott, A. J. (2000) *The Cultural Economy of Cities*, London: Sage.
Sennett, R. (1986) *The Fall of Public Man*, London: Faber and Faber.
—— (1990) *The Conscience of the Eye*, London: Norton.
—— (1994) *Flesh and Stone*, London: Faber and Faber.
—— (1996a) 'The Social Body', *Transition*, 71, 80–98.
—— (1996b) *The Uses of Disorder*, London: Faber and Faber.
—— (1998) 'The Sense of Touch', *Architectural Design*, 68 (3/4), 19–22.
Sharp J., Pollock, V. and Paddison, R. (2005) 'Just Art for a Just City: Public Art and Social Inclusion in Urban Regeneration', *Urban Studies,* 42 (5/6), 1001–23.
Shearing, C. and Stenning, P. (1996) 'From the Panopticon to Disney World: The Development of Discipline', in E. Muncie, E. McLaughin and M. Unger (eds) *Criminological Perspectives*, London: Sage.
Sheller, M. and Urry, J. (2003) 'Mobile Transformations of "Public" and "Private" Life', *Theory, Culture and Society*, 20 (3), 107–25.
Shields, R. (1989) 'Social Spatialization and the Built Environment: The West Edmonton Mall', *Environmental Planning D*, 7, 147–64.
—— (1991) *Places on the Margin*, London: Routledge.
—— (1992a) 'A Truant Proximity: Presence and Absence in the Space of Modernity', *Environmental Planning D*, 10, 181–98.
—— (1992b) 'The Individual, Consumption Cultures and the Fate of Community', in R. Shields (ed.) *Lifestyle Shopping*, London: Routledge.
—— (1992c) 'Spaces for the Subject of Consumption', in R. Shields (ed.) *Lifestyle Shopping*, London: Routledge.
—— (1996) 'A Guide to Urban Representation and What to Do about It: Alternative Traditions to Urban Theory', in A. D. King (ed.) *Re-Presenting the City: Ethnicity, Capital, and Culture in the 21st-Century Metropolis*, Basingstoke: Macmillan.
—— (1997) 'Flow as a New Paradigm', *Space and Culture*, 1, 1–7.
—— (1999) *Lefebvre: Love and Struggle*, London: Routledge.
Sibley, D. (1995) *Geographies of Exclusion*, London: Routledge.
Simmel, G. (1959) 'The Ruin', in G. Wolff (ed.) *Georg Simmel*, Columbus, OH: Ohio State University Press.
—— (1971a) *On Individuality and Social Forms*, Chicago, IL: University of Chicago Press.
—— (1971b) 'The Metropolis and Mental Life', in D. N. Levine (ed.) *On Individuality and Social Forms: Selected Writings*, Chicago, IL: University of Chicago Press.
—— (1994) 'The Sociology of the Meal', trans. M. Symons, *Food and Foodways*, 5 (4), 345–50.
—— (1997) 'Sociology of the Senses', in F. Frisby and M. Featherstone (eds) *Simmel on Culture*, Sage: London.
Simonsen, K. (2005) 'Bodies, Sensations, Space and Time: The Contribution from Henri Lefebvre', *Geografiska Annaler*, 87 B (1), 1–14.
Slater, T. (2005) 'Gentrification in Canada's Cities: From Social Mix to "Social Tectonics"', in R. Atkinson and G. Bridge (eds) *Gentrification in Global Context*, London: Routledge.

Smith, A. (2005) 'Conceptualizing City Image Change: The "Re-Imaging" of Barcelona', *Tourism Geographies*, 7 (4) 398–423.

Smith, N. (1996) *New Urban Frontier: Gentrification and the Revanchist City*, London: Routledge.

—— (1999) 'Which New Urbanism? New York City and the Revanchist 1990s', in R. Beauregard and S. Body-Gendrot (eds) *The Urban Moment*, London: Sage.

—— (2000) 'Del Lower East Side al Raval', *La Vanguardia*, 8 December 2000.

—— (2002) 'New Globalism, New Urbanism: Gentrification as Global Urban Strategy', *Antipode*, 34 (3), 427–50.

Smyth, H. (1994) *Marketing the City*, London: Spon.

Soja, E. (1996) *Thirdspace*, Oxford, Blackwell.

—— (2000) *Postmetropolis*, Oxford, Blackwell.

Sole, C. (2008) 'La nueva inmigracion', in M. Degen and M. Garcia (eds) *La Metaciudad: Barcelona – transformación de una metropolis*, Barcelona: Anthropos.

Sorkin, M. (1992a) *Variations of a Theme Park*, New York: Noonday Press.

—— (1992b) 'Introduction: Variations on a Theme Park', in M. Sorkin (ed.) *Variations of a Theme Park*, New York: Noonday Press.

—— (1992c) 'See you in Disneyland', in M. Sorkin (ed.) *Variations of a Theme Park*, New York: Noonday Press.

Stallybrass, P. and White A. (1986) *The Politics and Poetics of Transgression*, London: Methuen.

Stevenson, N. (2005) 'Media, Cultural Citizenship and the Global Public Sphere', in R. D. Germain and M. Kenny (eds) *The Idea of Global Civil Society*, London: Routledge.

Stewart, K. (1988) 'Nostalgia – A Polemic', *Cultural Anthropology*, 3 (3), 227–42.

Stewart, L. (1995) 'Bodies, Visions, and Spatial Politics: A Review Essay on Henri Lefebvre's *The Production of Space*', *Environmental Planning D: Society and Space*, 13, 609–18.

Subirats, J. and Rius, J. (2005) *Del Xino al Raval: Cultura i transformacio social a la Barcelona central*, Barcelona: Centro de Cultura Contemporanea de Barcelona.

Sudjic, D. (1992) *The 100 Mile City*, London, André Deutsch.

Sullivan, T. and Gill, D. (1975) *If you Could See What I Hear*, New York: Harper and Row.

Sust, X. (1986) 'La reconversion de la vivienda antigua en El Raval', in *Barcelona, metropolis mediterranea: la rehabilitacion de Ciutat Vella*, Barcelona: Ajuntament de Barcelona.

Synnott, A. (1991) 'Puzzling over the Senses: From Plato to Marx', in D. Howes (ed.) *The Varieties of Sensory Experience*, Toronto: University of Toronto Press.

Talen, E. (1999) 'Sense of Community and Neighbourhood Form: An Assessment of the Social Doctrine of New Urbanism', *Urban Studies*, 36 (8), 1361–89.

Tatjer i Mir, M. and Costa i Riera, J. (1989) 'Grups socials, agents urbans: estrategies i conflictes a Ciutat Vella de Barcelona', in *Primeres jornades Ciutat Vella*, Barcelona: Ajuntament de Barcelona.

Tello i Robira, R. (1993) 'Barcelona post-Olimpica: de ciudad industrial a ciudad de consumo', *Estudios Geograficos*, 212, 507–19.

Till, K. (2005) *The New Berlin: Memory, Politics, Place*, Minnesota, MN: University of Minnesota Press.

Time Out (2001) *Barcelona*, London: Penguin.

Time Out (2004) *Barcelona*, London: Penguin.
—— (2006) *Barcelona*, London: Penguin.
Thomsen, C. W. (1998) *Sensuous Architecture: The Art of Erotic Building*, Munich: Prestel Verlag.
Thrift, N. (1996) *Spatial Formations*, London: Sage.
Tonkiss, F. (2005) *Space, the City and Social Theory*, Cambridge: Polity Press.
Tucker, A. (1997) *Madola, esculturas – broshure*, Manchester: Manchester International Arts.
Union Temporal d'Escribes (UTE) (2004) *Barcelona, marca registrada*, Barcelona: Virus Editorial.
Urban Task Force (1999) *Towards an Urban Renaissance*, London: Urban Task Force.
Urry, J. (1990) *The Tourist Gaze*, London: Sage.
—— (1995) *The Consumption of Place*, London: Routledge.
—— (1999) 'Sensing the City', in D. Judd and S. Fainstein (eds) *The Tourist City*, London: Yale University Press.
—— (2000) *Sociology Beyond Societies*, London: Routledge.
—— (2002) *The Tourist Gaze*, 2nd edn, London: Sage.
Villar, P. (1996) *Historia y leyenda del Barrio Chino 1900–1992*, Barcelona: Ediciones La Campana.
Virilio, P. (1997) 'The Overexposed City', in N. Leach (ed.) *Rethinking Architecture*, London: Routledge.
Walker, I. (1993) 'Tourism Development and Tourism Policy in Manchester: A Case Study in Urban Tourism', unpublished MA dissertation, University of Lancaster.
Ward, K. (1998) *Selling Places: The Marketing and Promotion of Towns and Cities 1850–2000*, London: Spon.
—— (2000) 'Front Rentiers to Rantiers: "Active Entrepreneurs", "Structural Speculators" and the Politics of Marketing the City', *Urban Studies*, 37 (7), 1093–1107.
Weintraub, J. (1997) 'The Theory and Politics of the Public/Private Distinction', in J. Weintraub and K. Kumar (eds) *Public and Private in Thought and Practice*, Chicago, IL: University of Chicago Press.
Whatmore, S. (2001) *Hybrid Geographies*, London: Sage.
Williams, G. (2003) *The Enterprising City Centre: Manchester's Development Challenge*, London: Spon.
Williams, R. (1965) *The Long Revolution*, London: Pelican.
—— (1975) *The Country and the City in the Modern Novel*, Oxford: Oxford University Press.
Wilson, D. and Grammenos, D. (2005) 'Gentrification, Discourse and the Body: Chicago's Humboldt Park', *Environment and Planning D*, 23, 295–312.
Wilson, D., Wouters, J. and Grammenos, D. (2004) 'Successful Protect-Community Discourse: Spatiality and Politics in Chicago's Pilsen Neighborhood', *Environment and Planning A*, 36, 1173–90.
Wilson, E. (1991) *The Sphinx in the City*, London: Virago Press.
Wirth, L. (1995) [1938] 'Urbanism as a Way of Life', in P. Kasinitz (ed.) *Metropolis: Centre and Symbol of our Time*, Basingstoke: Macmillan.
Wycherley, R. E. (1962) *How the Greeks Built Cities*, London, Macmillan.
Wynne, D. and O'Connor, J. (1998) 'Consumption and the Postmodern City', *Urban Studies*, 35, 841–64.

Young, I. (1990) 'The Ideal of Community and the Politics of Difference', in L. Nicholson (ed.) *Feminism/Postmodernism*, London: Routledge.

Zucker, P. (1966) *Town and Square: From the Agora to the Village Green*, New York: Columbia University Press.

Zukin, S. (1988) [1982] *Loft Living: Culture and Capital in Urban Change*, London: Radius.

—— (1991) *Landscapes of Power: From Detroit to Disney World*, Berkeley, CA: University of California Press.

—— (1995) *The Cultures of Cities*, Oxford: Blackwell.

—— (1998a) 'Urban Lifestyles: Diversity and Standardisation in Spaces of Consumption', *Urban Studies*, 35, 825–39.

—— (1998b) 'Politics and the Aesthetics of Public Space: the American Model', in P. Subiros (ed.) *Ciutat Real, Ciutat Ideal*, Barcelona: Centre de Cultura Contemporanea de Barcelona.

Index

Note: page numbers in **bold** indicate figures and tables.

absence, and power 61–2, 116; spaces of 119
accessibility, Castlefield 106–9, 119, 130; economy of access 20–2, 195; El Raval 109–12, 119, 130; spatial technique 106
actor-network theory 57
aestheticization 9, 27–8, 36–8
aesthetics 8–9, 14, 16–17, 25, 26–30, 36–8; *see also* 'socially embedded aesthetic'; urban aesthetic
'affordances' 48, 49; *see also place gestures*
Allen, John 60–1, 71
appropriation of space, El Raval 170–2, **171**, **172**, 181–2
architecture, Modernist 11, 94

Baltimore harbour redevelopment 8
Barcelona, Catalan capital 94; Centre of Contemporary Culture 98; Cerdá's 'Eixample' 80, 93–4, 97; Franco period 80–1; Museum of Contemporary Art of Barcelona *see* MACBA; Olympic Games 94–5, 96, 97; the Ramblas 97, 143; Royal Gold Medal for Architecture 77; urban reconstruction (1979–85) 95–6
Barcelona regeneration 8, 14, 29, 94–102; 'Barcelonity' 97; Barceloneta 97; neighbourhood associatons 96, 98; Old City 16, 29, 190; social problems 98, 190
Bauman, Z. 45, 119–20, 141
Benjamin, Walter 38, 39, 146; on the modern city 38–9, 52
Bernstein, Sir Howard 92
Bohigas, Oriol 95, 97, 157
Bourdieu, P., *Distinction* 179
Brewer, J. 23
Burawoy, M. 12; *Global Ethnography* 12

Canary Wharf 87
Castlefield **6**, 35, 78, **122**; industrial wasteland 79, 83, 133, 145; history 78–83; industrial decline 145–6; map **85**; negative reputation 78–9, 86, 124; original character 78–9, 133–5; rediscovery of 83–4
Castlefield regeneration 5–6, 8, 83–93; accessibility 106–9, 136; brand name 135, 136; café-bar culture 88, 89, 92, 124, 164–5; Castlefield Information Centre 90, 91, 107; Castlefield Management Centre 90, 124; Catalan Art Festival (1995) 88; Catalan Square 89; CMDC (Central Manchester Development Corporation) 87–91, 106; conservation group 83; detachment from city centre 137; drinking area 124, 176, 193; Event Arena 90, **115**, 122, 168–9; events organized 122, 123–4, **123**; everyday rhythms 163–6, 173–6; exclusive residential area 137–9; flagship for Manchester 93, 104, 124; Granada Studios 85, 136;

heritage trail 121; history manufactured/commercialized 145, **145**, 147, 150; leisurization 121, 172–3, 175; loss of atmosphere 146–8; Manchester City Council Plan 91; map **84**; market 122–3, 163–4, **164**, 177–8, **178**; Mediterranean lifestyle theme 89, 121, 123, 164–5, **166**; Merchants Bridge 107, **108**; middle class development 108–9, 123, 136–7, 143; mixed use development 86, 91, 177; Museum of Science and Industry 85, 135, 163; negative reputation 78–9, 86, 124; Outstanding Conservation Area 83, 106; overview **103**; property market 136; public life 120–5, 176–9; public space, creation of 93, 107, 109; radical makeover 84–7; Roman fort and gardens 85–6, 135, 147, **148**; selective histories 144–51; sensescapes 89, 164–5; sensory map 174–5, **174**; Slate Wharf 145; spatial contestations 121–2, 176–8; 'taste wars' 176–9, 190, 197; 'third' space of interaction 172; Tourism Development Plan 84; tourist destination 87, 88; tourists' views 150–1; urban transformation 4–6, 14, 18, 35–6, 86–9; Urban Heritage Park 84, 86–7, 90, **93**, **107**, **108**, 112, 121, 135, 137, 145, 150, 163, 164, 166; Urban Rangers Service 90; Victorian railway viaducts **114**, 168, **170**; visual sense 87, 147–8
Castlefield residents 17, 18, 93, 133–5, 137–8, 145–50, 177; appropriation of space 181–2; imagination 148–9, 151, 196; 'lived' perceptions 136–7; local needs 135, 135–6, 177, 197; memories 145–8; ownership 190–1; user resistance/subversion 161, 192–3, 196
Castlefield and El Raval: absence of sensuous contrast 114–17; accessibility 106–12, 161; culture-led regeneration 6, 104; designer heritage aesthetic 106, 112–18, 161; differentiating features 112, 114, 117–19, 120–1; ethics of engagement 82, 167–8; fear of the 'Other' 191; global tourism and investment 161; obliteration of past 116–17; radical inversion 112–13, 114; re-use of vernacular architecture 116–19; sensescapes of decay 82–3; sensuous mappings 173–6
Central Manchester Development Corporation (CMDC) 87–91, 106
Cerdá, Ildefons, 'Eixample' 80, 93–4, 97, 110
Certeau, M. de 55, 64, 133
cities 7, 11, 12, 16, 54; as conceived and as lived 18, 55–6, 132–3; fear and safety 69–72; global competition 17, 26–7, 86, 116, 117, 197; global network 7; key players in world economy 7, 11, 26–7; 'layered city' 182; revival of 16; planning 65–73; as product 26, 27, 34; and rhythms 50–2, 162–76; social cohesion 9, 68, 70, 166, 196; symbolic economy 27, 29; UK government policy 87; unique character 12, 29, 30; visual aspect 36–7; *see also* urban planning; urban regeneration
city of senses 41–7
civility 22–3, 126–7, 141, 181
Clifford, J. 13
consumption of place 31–2, 33, 34, 87, 121–4
control society 60–2, 73; power techniques 59–60; through pleasure 60, 70–1
Corbin, A. 66–7
'cultural re-coding' 28, 86
culture, urban development 4–5, 8–9, 27–30, 98–9

Davis, M. 33, 70, 71–2, 112
designer heritage aesthetic 71, 106, 112–18, 130, 133, 150, 161, 169, 191, 195
discipline in society 59, 60
Disney World 60
dominated space 62–4, 116, 172
Douglas, Mary 67–8
Dovey, K.105

economy of access 20–2, 106–12, 195, 197; *see also* accessibility

Edensor, T. 26, 116, 145, 199
El Raval **5**, 35, 78, 80; after dark 179–81; authentic neighbourhood 150, 186; balconies, significance of 117–18, 119, 158–9, **159**, 198; 'Barrio Chino' 80–2, 139–40; Cerdá's 'Eixample' 80, 93–4; drug culture 82, 83, 140; exodus of residents 1960s 82; history 78–83; immigration 6, 80, 126, 186–91, 198–9; negative reputation 80–1, 83, 96, 101, 133, 139–40; north–south divide 80, 175; planning history 93–102; prostitution/sex related industry 81, 101, 109; smells 167; social deprivation 83, 96, 98; social life and networks 167, 182–4, **183**; *see also* Barcelona
El Raval, old and new contrast 17–18, **118**; MACBA/old buildings 155, **156**; old stationery shop **185**; old streets **120**, 167, 168; Plaça dels Angels (1900 and 1999) **113**, **115**; regenerated street **120**
El Raval regeneration 197–8; accessibility 109, 111, **111**, 112; appropriation of space 170–2, **171**, **172**, 181–2, 196; civic consensus 127; civilizing process 110, 126–7; commercial value 112; contestation of space 179–82; contradictory representations 143–4; cultural quarter 102, 128–9, 140, 142, 195; demolition 99, 101; design 96, 99, 158–9; destruction of living social history 151–2, **152**, 156; dilution/displacement of existing population 126–8, 195; distinctive character 184–6; 'dwelling' place 175; ethics of engagement 167; expulsion of residents 101, 126, 143–4, 129–30; gentrification 128–9; homogenization 97, 125, 130, 142, 184, 195; invisible public places 80, 143; MACBA 98, 102, 112, 114; map **94**, **100**; multicultural character 102, 186–91; neighbourhood associations 98, 101; non-European immigration 102, 186–91, 193; cultural regeneration 98–9; overview **103**; PERI, interior reform plan 96; 'place wars' 179–81, 190, 197; planners' vision 93–102, 106–20, 125–30; poverty/marginality 102, 104, 130, 186; private property developers 96, 126, 143, **144**; property prices 126, **144**; public space 93–102, 112, 125, 139–44, 173, 175; purification of space 110–12; Rambla del Raval 99, 127; *ravalejar* 191, **192**; rhythm of activities 166–72; sensuous mapping 167, **174**, 175; 'social mix' 125–30; social problems 102, 127, 129–30, 198; southern Raval, economic and social regeneration 98; urban transformation 4–6, 14, 18, 35–6, 101; zoned spatialization 168
El Raval residents 13; appropriation of space 170–2, **171**, **172**, 181–2; associations 152–3, 155, 156; celebrate being 'different' 184–5; character of neighbourhood 155, 158; demolition of neighbourhood 151–2, 154, **154**, 156; displacement of 143–4; expectations of 139–40; heroin problem as scapegoat 140; history of neglect by authorities 153–5, 185; homogenization resisted 142, 184; immigrants as scapegoats 141, 189, 191, 197; lack of official consultation 155, 157; multicultural character 189, 190; negative reputation of area 139–40, 155; perceive need for change 160–1; positive reactions 140, 142, 153; real needs ignored 158; rehousing 157, 158–9; self identity 133, 139; sense of place lost 156–7, 196; social life/networks, damaged **154**, 155–6, 158–9, 160, 184; threat of minorities 141; tourism versus residents 112, 141–3, 167, 197–8; transformation not accomplished 140–2
Ellin, N. 69, 71
ethics of engagement 22–4, 87, 166, 167, 193, 195, 197–8
exclusion by design 32, 68–70, 194–200

fear and the urban environment 69, 70–1, 79, 92, 141, 191; and fortification of urban space 30–4, 139, 179
Featherstone, M. 27, 28

flagship squares 168–73; Castlefield Event Arena **115**; Plaça dels Angels **115**
'fortress city' 33–4
Foucault, M.: on discipline 14, 59; on power 59–60, 61

gated developments 33, 139, 179
GATPAC, group of modernist architects 94
gentrification 9; El Raval 127–30
Gibson, James 48
globalization 7–8, 23; global property market 16, 143
Gordon, A. 49
graffiti 54; El Raval 157, **157**, 179, **180**
Grosz, E. 48–9

Habermas, J. 24
Harvey, D. 8, 25–6, 29; 200; 29
Haussman, Georges Eugène, Baron 66
Hayden, D.152
Heidegger, M. 'Building, Dwelling, Thinking' 21
heritage *see* designer heritage aesthetic
Hetherington, K. 57
Howes, D. 10, 37–8
human/material relations 47–50, 57–8

Jacobs, Jane 21, 51
Jacobs, J. M. 109, 110
Jay, M. 47

Lash, S. 27–8
Latour, B. on power 57–8
Lefebvre, H.: on power 56, 58; rhythmanalysis 15, 50–3, 62, 173; on space 10, 13, 16, 50, 62–3, 162, 194; total body 13, 15, 50, 161, 194; triad of social space 18–20, 104, 105, 195
Levinas, E. 23
Levine, M. V. 8
Lingis, A. 40
lived experience *see* urban regeneration
local populations *see* urban regeneration
London docks regeneration 8
Los Angeles, redevelopment 32, 33
Lowell's National Historic Park, Connecticut 84
Lukes, S. 61
Lynch, K. 41
Lyon, D. 34

MACBA, Museum of Contemporary Art of Barcelona 98, 102, 112, 114, 150, 155, 168, **168**, **169**
Maffesoli, M. 32
Manchester city centre regeneration 8, 14, 16, 77, 88, 91–3, 132; and Barcelona 77, 88; city centre population loss 83; European city image 88, 89, **90**, 92; IRA bomb (1996) 91; social deprivation and crime 90; *see also* Castlefield regeneration
Marcuse, P. 33
Marcuse, P. and Kempen, R. van 182
marginal areas, lack of investment 102, 153, 155
Massey, D. 7, 196
Meier, Richard 98
Merleau-Ponty, M. 39
Merman, M. 69
middle class public 108–9, 123, 136–7, 143, 195
Miles, M. 28
Mitchell, D. 24
Mitterrand, François 63
monumental spaces 62, 64
multiculturalism 187–91, 199

Newman, Oscar *Defensible Space* 70

O'Connor, J. and Wynne, D. 89, 108
'other' 69, 71, 191

Paris, Grand Arche, la Défence 63, 64
PERI, Interior Reform Plans, Barcelona 96
phenomenology 39–41, 48
Pile, S.10, 65, 199
Plaça dels Angels; at night 179–81; events organized 167, 181, **181**; local appropriation of space 170–2, **171**, **172**; new immigrants 170, 172; rhythm of activity 170–2
place 7; commercialization of 146; 'cultural re-coding' 28–9; experience of 14–15, 17, 26–7, 60; identities 29, 72; *see also* space

place gestures 47–50, 53, 62, 148, 161
politics of representation 24–5, 197–8; *see also* public life; public space
Porteous, J. D. 42
power 54; absence, role of 61–2, 116; ambient 60–1, 71; built environment 55–6, 73; control society 60, 62, 73; control through pleasure 60, 70–1; discipline 14, 59–60; discourse of colonization 110; domination and resistance 55, 62–5, 161; fluid and infiltrating 14, 55–62, 72; medicalization of 99; modalities of 55, 58, 60–2; monumental spaces 62, 64; network of associations 55, 73; of place 152; post-panopticon stage 59–60; relational 14, 55, 57–9, 72; sensuously mediated 14, 54, 57, 61; techniques 59, 60
public life 7, 10–11, 193, 194; commercially regulated forms 17, 31–3; consumption experience 30–1, 92, 124; as envisaged 14; economy of access 20–2, 106–12, 195, 197; ethics of engagement 22–4, 87, 166, 167, 193, 195, 197–8; dilution of 125–30; formalization of 121, 122–5; loss of social diversity 31, 32; politics of representation 24–5, 197–8; and senses 20–25; *see also* sociality
public space 3–4, 7, 11–12, 16, 143, 194, 198–9; a battlefield 9, 190, 193; Castlefield 93, 107, 109; cleanliness 67–9; commercial aspect 31, 106; conceived vision 18; economy of access 20–2, 106–12, 195, 197 ; El Raval 112, 125; 'end of public space' thesis 10–12; ethics of engagement 22–4, 193, 197–8; exclusion 106; expression of power 14; fear, control of 69–72; fluid character of 12, 20–2, 65; fortification of 33–4; image promotion 30; invisible 143; 'lived localization' El Raval 173, 175; physical reorganization 14, 16; political dimension 24–5; politics of representation 24–5, 197–8; as product 31, 92, 194; social dimension 4, 22–4; and purification 86–7, 99, 110–12, 195; social order 66–7; *see also* social space; space
'publicness' 20–25, 112, 190, 194
purity/impurity and space 67–9, 99

Quilley, S. 89

racialization of space 172, 189–91
Ramsbottom, Jim 91
resistance/domination 63–5, 73, 161
rhythmanalysis 50–3, 162–76
Rifkin, J. 33, 139
Rodaway, P. 48, 51–2

Sassen, Saskia 7
Sennett, Richard 10–11, 21, 22, 22–3, 33, 37, 43, 54, 56, 59, 72, 159, 193
senses, and body 3–4, 9, 18, 162; domination and resistance 62–5; framing social relations 3–4, 39, 56; framing urban planning 65–73, 77, 194; and spatial practices 77; and order 119; power relations in built environment 54, 55–6, 57, 59, 72; 'publicness' 21–2, 25, 194; researching the senses 12–13; shaped by ideologies in society 72; urban experience 7, 38–9, 73
'sensescapes' 42, 48, 51–2, 89, 105, 164–5
sensory experience 3–4, 12, 35–41, 72; absence of 61–2; and city 41–7; historical forms 52; pollution 79; urban space 9, 17, 36, 41, 178
sensory mappings 162, 173–6
sensuous encounters 3; and aesthetic, social exclusion 130; engagements, El Raval 167; feeling of loss, Castlefield 146–8; geographies 67, 81, 105, 130, 157, 175; ideologies 55, 65–72, 104, 112, 125
Sheller, M. and Urry, J. 20
Shields, R. 20, 32
Sibley, D. 68, 110, 140, 141
Simmel, George 22, 39, 149; on the senses and the city 38–9, 47, 52
smell, sense of 66–7; city of smell 44–5
Smith, Neil 109, 127, 128, 130–1
social exclusion 127, 129–30

'social mix' 125–30
social space, Lefebvre 10, 13, 15, 18–20, 62–3, 116, 162, 172; appropriated space 63; conceived/designed space 19, 25, 55–6, 72; dominated space 62–3, 64, 116, 172; lived/experienced space 19, 25, 39, 55–6, 72; monumental space 62, 64; perceived *spatial practices* 18, 25; *see also* appropriation of space, El Raval; space
sociality/sociability patterns 16–18, 30, 32–4, 56
'socially embedded aesthetics' 10, 14, 38, 40, 50, 52, 56, 104, 195–6; and lived space 191; analysis of power relations 72; analysis of public space 65, 67; of place 50, 162
Soja, E. 19, 25, 162
Sony Centre, Berlin 60
sound, city of sound 43–4
space, abstract 63; homogenized 71, 72; 'networks' 57; phenomenological approach 39–41; public/private 20; as social process 17–20; 'third' space of interaction 172; urban design 26; *see also* appropriation of space, El Raval; social space
spatial order in the city 73, 80, 101
Stallybrass, P. and White, A. 69, 144
Subirats, J. and Rius, J. (2005) 186
surveillance 33–4, 70, 92

taste, city of taste 45–6
Thatcher, Margaret, UK Prime Minister 87
Times Square, New York 29, 31
touch, city of touch 42–3

'unsafety', Bauman 191
urban aesthetic 16–17, 25, 26–30, 36–8
urban design 16, 25–30, 96, 99, 158–9
Urban Development Corporations (UDC) 87
urban planning 9, 25–6, 73; conceived vision 14, 18–19; control through pleasure 8, 70; doctors of space 99, 129–30; global strategies 130–1; ideologies 66–72, 73; security, significance of 70; social differentiation and discrimination 33–4; use of history 8, 70
urban regeneration 6–12, 25, 30–1, 58, 130; culture 4–5, 8–9, 27–30; lived experience of 10, 15, 18, 19; local populations 8, 9, 17–18, 65, 135, 161, 193; and neo-liberal politics 9, 127; new sensorial order 119; policy 87; recommendations for the future 194, 199–200; schemes 6, 53; social improvements 129–30; social inequalities 29–30; spatial techniques 64, 104, 106, 120; theming 29; visual dimension 36–7; *see also* public space
Urry, J. 48, 151; *see also* Sheller, M. and Urry, J.
users, multiple experiences and meanings 151, 163, 192–3, 196

vernacular post modernism 117
vision, city of vision 42, 46–7
visual, hegemony 87, 147–8, 161; symbols 168–9

White, A. *see* Stallybrass, P. and White, A.
Williams, R. 150
Wynne, D. *see* O'Connor, J. and Wynne, D.

zoning, city development 69, 70, 79
Zukin, S., 27, 29, 36–7, 60, 92, 115, 116, 129, 181